ESTIMATING THE PARAMETERS OF THE MARKOV PROBABILITY MODEL FROM AGGREGATE TIME SERIES DATA

CONTRIBUTIONS
TO
ECONOMIC ANALYSIS

65

Edited by

J. JOHNSTON

J. SANDEE

R. H. STROTZ

J. TINBERGEN

P. J. VERDOORN

NORTH-HOLLAND PUBLISHING COMPANY
AMSTERDAM · LONDON

ESTIMATING THE PARAMETERS OF THE MARKOV PROBABILITY MODEL FROM AGGREGATE TIME SERIES DATA

T. C. LEE
University of Connecticut

G. G. JUDGE
University of Illinois

A. ZELLNER
University of Chicago

1970

NORTH-HOLLAND PUBLISHING COMPANY
AMSTERDAM · LONDON

Publishers:
NORTH-HOLLAND PUBLISHING COMPANY – AMSTERDAM
NORTH-HOLLAND PUBLISHING COMPANY, LTD. – LONDON

PRINTED IN GERMANY

INTRODUCTION TO THE SERIES

This series consists of a number of hitherto unpublished studies, which are introduced by the editors in the belief that they represent fresh contributions to economic science.

The term *economic analysis* as used in the title of the series has been adopted because it covers both the activities of the theoretical economist and the research worker.

Although the analytical methods used by the various contributors are not the same, they are nevertheless conditioned by the common origin of their studies, namely theoretical problems encountered in practical research. Since for this reason, business cycle research and national accounting, research work on behalf of economic policy, and problems of planning are the main sources of the subjects dealt with, they necessarily determine the manner of approach adopted by the authors. Their methods tend to be 'practical' in the sense of not being too far remote from application to actual economic conditions. In addition they are quantitative rather than qualitative.

It is the hope of the editors that the publication of these studies will help to stimulate the exchange of scientific information and to reinforce international cooperation in the field of economics.

THE EDITORS

PREFACE

A Markov chain model is one appropriate probability model for time series data when the observation at any point in time is the state or category into which the unit being observed falls. Although the Markov probability model is of relatively recent vintage, it has been used as a basis for describing the generation of sets of data over a wide variety of subject matter areas. Methods of statistical inference for the parameters of the probability system, when time ordered micro data are available, are of long standing and application of these methods are well reported in the literature. However, consideration of the problem of estimating the transition probabilities when the data are generated by a first order Markov process and only aggregate outcome data are available, dates back only to the beginning of the 1950's and to a large extent, the major work has occurred during the last five years.

Given this situation, this book is concerned with (1) summarizing and evaluating the initial results which are scattered throughout the literature, (2) developing alternative macro transition probability estimators and the corresponding computer routines and (3) evaluating the finite sample properties of these various estimators by a limited sampling experiment. Although the results reported apply primarily to aggregate data generated from a stationary first order Markov process, the extension of the results to areas concerned with the estimation of transition probabilities that are not time constant and the general problem of estimation when proportion data are used are considered in appendices.

Many persons helped to shape the final form of this book. Articles by Miller (1952), Goodman (1953), Madansky (1959) and Telser (1963) pointed the way to the literature in this field and provided valuable insights as to how the estimation problem might be solved. Takashi Takayama made major contributions to the first stages of the work and has been a valuable critic of the various drafts of the book.

7

Mary Ellen Bock helped in developing computer programs, read several versions of the manuscript and did much to make the book more readable and logically complete. R. Cain helped in developing computer programs, generated the sample data and worked on some of the early analyses. William E. Griffiths and Don H. Miller read the manuscript at a last rough draft stage and made many valuable comments. Carol Daggett rapidly and accurately typed the various drafts of the manuscript, caught many errors and contributed much to the book's final format. To these and to many others, we say thanks for being such effective colleagues.

T. C. Lee
G. G. Judge
A. Zellner

CONTENTS

INTRODUCTION TO THE SERIES 5

PREFACE 7

LIST OF SYMBOLS 15

Chapter 1

INTRODUCTION 17
   1.1. The Markov probability model 18
   1.2. Estimation problems 20
   1.3. Plan of the book 21

Chapter 2

THE ESTIMATION OF TRANSITION PROBABILITIES FROM MICRO DATA 23
   2.1. The micro maximum likelihood (ML) estimator 23
   2.2. Bayesian analysis of the micro model 25
      a. Bayes' theorem and Bayesian inference 26
      b. The prior density function 27
      c. The posterior probability density function 29
      d. Bayesian estimation 30

Chapter 3

THE ESTIMATION OF TRANSITION PROBABILITIES FROM MACRO DATA 31
   3.1. A relation involving the macro data 31
   3.2. The unrestricted least squares transition probability estimator . . 32
      a. The row sum condition 34
      b. The $0 \leqq p_{ij} \leqq 1$ condition 36
      c. Some properties of the unrestricted estimator $\tilde{p}$ 38
   3.3. The restricted least squares transition probability estimator . . . 39
      a. Sampling properties of the restricted estimator 41

Chapter 4

THE SAMPLING EXPERIMENT AND SOME INITIAL RESULTS 43
4.1. The simulated probability model and its characteristics . . . 44
4.2. The procedure of simulation 46
4.3. The simulated population and sampling 46
4.4. Sample proportions as the estimates of true proportions 47
4.5. Basis for gauging estimator performance 49
4.6 Experimental results for the maximum likelihood estimator using micro data 51
4.7. Results from generated probability vector time series data . . . 54
4.8. Results from the sampling experiment macro data 56
a. Unrestricted least squares 56
b. Restricted least squares 57
4.9. An application 59
a. The brand change problem 59
b. Results for the restricted and unrestricted estimators 60

Chapter 5

WEIGHTED INEQUALITY RESTRICTED LEAST SQUARES ESTIMATORS 63
5.1. The statistical model 64
5.2. Weighted restricted least squares 64
5.3. Some alternative weights 66
5.4. Results from sampling experiment 67
a. Weighted by average proportion of ith state 67
b. Weighted by an estimate of the ith equation disturbance variance . 68
c. Weighted by product of average proportions in state i 68
5.5. Results for the brand change problem 71

Chapter 6

A GENERALIZED LEAST SQUARES ESTIMATOR 73
6.1. Non-spherical disturbances 73
6.2. Redundant parameters and the reduced model 75
6.3. Existence of the inverse of the disturbance covariance matrix . . 76
6.4. Aitken's generalized unrestricted and restricted least squares estimators 79
6.5. Results from the sampling experiment 81

Chapter 7

THE MINIMUM CHI-SQUARE ESTIMATOR 85
7.1. Preliminaries 85
7.2. The restricted minimum chi-square estimator 86

7.3. The modified restricted minimum chi-square estimator 87
7.4. An equivalent model 88
7.5. A numerical example 90

Chapter 8

THE MACRO MAXIMUM LIKELIHOOD ESTIMATOR 93
8.1. The multinomial distribution under the Lexis scheme 93
8.2. The mode of the likelihood function 95
8.3. The macro maximum likelihood estimator 99
8.4. Results from the sampling experiment 102
8.5. Some applications 102
   a. Results for the brand change problem 102
   b. Results for the tenure status problem 105

Chapter 9

BAYESIAN ANALYSIS OF THE 'MACRO' MODEL 107
9.1. Bayesian analysis: prior distribution 107
9.2. Likelihood function and posterior distribution 109
9.3. The mode of the posterior distribution 110
9.4. Comparison with some sampling theory results 112
9.5. A macro Bayesian transition probability estimator 115
9.6. The Bayesian approach: further considerations 117
9.7. A numerical example 119
9.8. Sampling experiment results 120
   a. Multi-beta leptokurtic prior 121
   b. Multivariate beta leptokurtic prior 123
   c. Results from a platykurtic prior 125
9.9. Results for the brand change problem 127

Chapter 10

THE MINIMUM ABSOLUTE DEVIATIONS ESTIMATOR 131
10.1. Specification of the statistical model 131
10.2. Linear programming formulation 132
   a. Minimizing the unweighted sum of the absolute deviations . . . 132
   b. Minimizing the weighted sum of the absolute deviations . . . 133
10.3. Results from the sampling experiment 134

Chapter 11

PREDICTION AND THE CHI-SQUARE GOODNESS-OF-FIT TEST 139
11.1. Predicted proportions 139
11.2. The chi-square goodness-of-fit test 141
11.3. Results from sampling experiment 142

Chapter 12

COMPARISONS OF THE ALTERNATIVE ESTIMATORS 145
12.1. The basis for comparison 145
12.2. Aggregate mean square error and variance measure 145
12.3. Wilcoxon matched-pairs signed-ranks test and Kendall's coefficient
   of concordance 150
12.4. Summary 156

Chapter 13

CONCLUDING REMARKS 157

Appendix A

THE GENERALIZED INVERSE METHOD 163
A.1. A generalization of the generalized least squares 163
A.2. The generalized inverse of the disturbance covariance matrix for the
   Markov probability model 168
A.3. The multicollinearity case 173
A.4. Row sum condition and the reduced weight matrix 174
A.5. The unique solution of the generalized inverse estimator is the Aitken's
   generalized least squares with redundant parameters deleted . . . 178
A.6. Summary 179

Appendix B

THE GENERAL LINEAR PROBABILITY MODEL 183
B.1. The model 183
B.2. The unrestricted estimator 185
B.3. The restricted estimator 186
B.4. A joint estimation procedure 190

Appendix C

ESTIMATION OF VARIABLE TRANSITION PROBABILITIES 191
C.1. The model 191
C.2. The unrestricted estimator 195
C.3. The restricted estimator 196
C.4. Concluding remarks 197

Appendix D

THE FORTRAN PROGRAMMING OF CLASSICAL AND BAYESIAN TRANSITION PROBA-
BILITY ESTIMATORS 199
D.1. The standard procedure 199

D.2. Incorporation of the prior knowledge with sample observations . . 200
D.3. Deleting a column 200
D.4. Assignment of weight 201
D.5. Recursive solutions for the ML and Bayesian estimators 201
D.6. The use of the control 'DITTO' and 'DIT.' 202
D.7. The use of the control 'CLEAR' and 'SUMMARY' 202
D.8. Option control card 203
D.9. An input example 205
D.10. An output example 206
D.11. Fortran listing of computer routine 209

REFERENCES 243

AUTHOR INDEX 251

SUBJECT INDEX 253

LIST OF SYMBOLS

Symbol	Meaning	First used and defined on page
a_{ij}	Prior parameters	28
α	Vector of slack variables for the primal problem	41
β	Vector of slack variables of the dual problem	41
$\tilde{\delta}$	Variable transition probability estimator	195
η_r	$(r \times 1)$ column vector with all elements 1	33
η_T	$(T \times 1)$ column vector with all elements 1	33
g_*	General linear probability estimator	185
g_*^c	Restricted general linear probability estimator	187
G	$(r \times r^2)$ matrix with r identity submatrices I_r	34
Γ	Gamma function	28
H	$(rT \times rT)$ weight matrix for a set of regression equations	64
$H'H$	$(rT \times rT)$ weight matrix for a set of normal equations	64
I_r	$(r \times r)$ identity matrix	34
J	$(r(r-1) \times r(r-1))$ diagonal matrix	111
λ	Vector of Lagrangean multipliers	34
Λ	Matrix of characteristic roots	165
$n_j(t)$	Number of micro units in state j at time t	23
$N(t)$	Total number of micro units at time t	48
Ω	Admissible region of parameter space	26
p	Column vector with subvectors p_j, $j = 1, 2, \ldots, r$	33
p_0	Prior mean of p_*	113
p_*	Column vector with subvectors p_j, $j = 1, 2, \ldots, r-1$	76
p_j	A vector obtained from jth column of P	32
p_{ij}	Transition probability (element of P)	19
$\dot{p}$	Micro maximum likelihood estimator	25
$\ddot{p}$	Micro Bayesian estimator	30
$\tilde{p}$	Unrestricted classical least squares	33
$\tilde{\dot{p}}_*$	Unrestricted Bayesian estimator	112
$\tilde{\dot{p}}_*^c$	Restricted Bayesian estimator	116
$\tilde{p}^c$	Restricted classical least squares	40
$\tilde{\tilde{p}}^1$	Restricted least squares weighted by average proportion	71
$\hat{p}_*$	Unrestricted generalized least squares estimator, or unrestricted minimum chi-square estimator, or unrestricted macro maximum likelihood estimator	79

Symbol	*Meaning*	*First used and defined on page*
$\hat{p}^c_*$	Restricted generalized least squares estimator, or restricted minimum chi-square estimator, or restricted macro maximum likelihood estimator	80
$\bar{p}_*$	Theil–Goldberger estimator	115
P	$(r \times r)$ transition matrix	19
π	Steady state probability vector	19
$q_j(t)$	True proportion for state j at time t	31
r	Number of Markov states	18
R	$(r \times (r - 1) r)$ matrix with $(r - 1)$ identity submatrices I_r	79
R_y	Sample space	26
Σ	Variance-covariance disturbance matrix for u	33
Σ_0	Prior variance–covariance matrix	111
Σ_*	Variance–covariance disturbance matrix for u_*	76
$\hat{\Sigma}_*$	Sample estimate of Σ_*	87
Σ^+	Generalized inverse of Σ	164
T	Number of time periods	18
θ	Parameter vector	26
u	Column vector with elements $u_j, j = 1, 2, \ldots, r$	33
u_*	Column vector with elements $u_j, j = 1, 2, \ldots, r - 1$	76
u_j	Column vector with elements $u_j(t), t = 1, 2, \ldots, T$	32
$u_j(t)$	Disturbance vector for state j at time t	32
x_t	Discrete random variable	18
X	$(rT \times r^2)$ matrix with r blocks of identical $X_j, j = 1, 2, \ldots, r$, on diagonal	33
X_*	$((r - 1) T \times (r - 1) r)$ matrix with $(r - 1)$ blocks of identical $X_j, j = 1, 2, \ldots, r - 1$, on diagonal	76
X_j	$(T \times r)$ matrix of observed proportions $y_i (t - 1)$. X_j's are identical for all	32
y	Column vector with subvectors $y_j, j = 1, 2, \ldots, r$	33
y_*	Column vector with subvectors $y_j, j = 1, 2, \ldots, r - 1$	76
y_j	Column vector with elements $y_j(t), t = 1, 2, \ldots, T$	32
$y_j(t)$	Observed proportion in state j at time t	32
$y_i (t - 1)$	Observed proportion in state i at time $t - 1$	32
Z	External explanatory variables	192

CHAPTER 1

INTRODUCTION

One of the objectives of positive economics is to explain or describe how outcome data associated with economic processes and institutions are generated. Knowledge of this nature is desired since it is generally assumed that if the processes underlying the attained values of economic variables are understood, then we shall be in a position to predict and/ or exercise some control over the future outcomes of the variables. Alternatively, such knowledge may suggest how the economic structure must be changed, or the values that instrument or control variables must take on, if certain target values or outcomes are to be achieved.

Within the context of economic theory, and in the words of Marschak (1950, p.3), economic data are usually viewed as being generated by a system of relations that is in general simultaneous, dynamic and stochastic. In addition, as Marschak (1953, p. 12) has further observed, few economic observations are free of errors and few economic relations are free of shocks. Therefore, the quantities that we want to predict are random variables and prediction consists of attaching probabilities to the various future possible outcomes or stating the probability distribution of these variables. Within the context of probability theory, a stochastic process is viewed by Cox and Miller (1965) as one that develops in time and/or space according to probabilistic laws. Therefore, the future outcome of the process is not deterministic and the best we may do is to attach probabilities to the various future outcomes or states. Given this type of probability model, the stochastic and dynamic properties of economic data have led many economists to postulate that the evolution of certain economic data may be characterized as a stochastic process. Within this context, the problem of predicting the future time path of economic variables becomes one of determining which stochastic process is associated with particular economic structures and proposed policy actions. In much theoretical and applied work to date, the

Markov chain process in discrete time is the main probability model which has been used in the analysis of economic time series where an observation at any given time is the category in which the observation falls.

A partial listing of research which makes use of the Markov model includes the work of Champernowne (1953) and Solow (1951) on income and wage distributions; the work of Prais (1955) and Goodman (1965) on social mobility; the work of Hart and Prais (1956), Adelman (1958), Preston and Bell (1961) and Steindl (1965) on the size distribution of firms and business concentration; the work of Sparks (1960) on analyzing consumer food purchases; the work of Gale (1960) and Smith (1961) in developing models for analyzing international and inter-regional trade flows; the work of Telser (1962, 1963) in analyzing consumer behavior; and the work of Cootner (1964) and others in analyzing stock price movements. Applications in other subject matter areas of the social sciences include such work as that of Bush and Mosteller (1955) on stochastic learning models; Blumen et al. (1955) and Matras (1960) on population and labor mobility; Coleman (1964) on diffusion of information; and Anderson (1955) on changes in voter attitude. In addition, it should be noted that Markovian decision processes constitute an important class of optimization models. Examples of work in this area are contained in research papers by Arrow (1967) and Shupp (1968) and a book by Howard (1960).

1.1. The Markov probability model

The type of stochastic process with which we will be concerned may be characterized as one in which (1) there are a finite number of possible outcomes s_i $(i = 1, 2, ..., r)$ which a discrete random variable $x_t (t = 0, 1, 2, ..., T)$ may take at a finite number of equidistant time points or trials t in a sample space where each elementary event is an infinite sequence $\{x_0, x_1, ...\}$; (2) the probability distribution of an outcome of a given trial depends only on the outcome of the immediately preceding trial and this first order dependence is the same at all stages, i.e.,

$$\Pr(x_t|x_{t-1}, x_{t-2}, ...) = \Pr(x_t|x_{t-1}), \qquad \text{for all } t, \qquad (1.1.1)$$

where $\Pr(x_t| \ldots)$ denotes the conditional probability density function for x_t. The probability of an ordered set of sequences is reflected by the multiplication law of conditional probability as

$$\Pr(x_0, x_1, \ldots, x_T) = \Pr(x_0)\Pr(x_1|x_0)\Pr((x_2|x_0)\, x_1) \ldots, \quad (1.1.2)$$

which may be written for a Markov process as

$$\Pr(x_0, x_1, \ldots, x_T) = \Pr(x_0)\prod_{t=1}^{T}\Pr(x_t|x_{t-1}). \quad (1.1.3)$$

Such a probabilistic mechanism or system is described by the initial probability distribution $\Pr(x_0)$ and the conditional probabilities $\Pr(x_t|x_{t-1})$. Further, it is assumed that if $x_{t-1} = s_i$ and $x_t = s_j$, then

$$\Pr(x_t = s_j|x_{t-1} = s_i) = p_{ij}(t) = p_{ij}, \quad \text{for all } t, \quad (1.1.4)$$

where p_{ij} is the *constant* transition probability associated with a change from state s_i to state s_j. These transition probabilities p_{ij} which may be arranged as an $(r \times r)$ transition probability matrix, $P = [p_{ij}]$, reflecting every pair of states, $s_i, s_j, (i, j = 1, 2, \ldots, r)$, have the following properties:

$$0 \leq p_{ij} \leq 1, \quad (1.1.5)$$

and

$$\sum_j p_{ij} = 1, \quad \text{for } i = 1, 2, \ldots, r. \quad (1.1.6)$$

Since the probabilistic behavior of a Markov chain is determined by the transition probability matrix P and a probability distribution over the initial state x_0, if we are given x_0 and P, we may want to determine the probability distribution for each random variable x_t or possibly we may be interested in the limiting distribution of x_t as $t \to \infty$, if such a distribution exists. Within this context, if a chain is irreducible and aperiodic and thus ergodic, then there exists a unique row vector $\pi = (\pi_1, \pi_2, \ldots, \pi_r)$, which is a steady state probability vector, with the properties

$$\lim_{t \to \infty} p_{ij}^{(t)} = \pi_j, \quad i, j = 1, 2, \ldots, r, \quad (1.1.7)$$

where $p_{ij}^{(t)}$ is the (i, j)th element of P^t,

$$0 \leq \pi_j \leq 1 \quad (1.1.8)$$

$$\sum_j \pi_j = 1, \quad (1.1.9)$$

and
$$\pi = \pi P. \qquad (1.1.10)$$

In this section, we have defined only those concepts needed as a basis for the chapters to follow. A more complete treatment of the theory of finite Markov chains is given by Kemeny and Snell (1960), Bailey (1964) and Cox and Miller (1965).

1.2. Estimation problems

When a sample of repeated observations of the chain exists, and time ordered data which reflect the intertemporal movements of the micro units over the states are available, then, as shown by Anderson and Goodman (1957), eq. (1.1.3) yields a likelihood function and serves as a basis for obtaining estimates of the transition probabilities and making certain tests of hypotheses about these parameters. Samples of time-ordered micro data and the availability of methods for estimation and inference formed the basis for the applications of the Markov model that were previously noted.

Unfortunately, data involving time-ordered detailed changes are frequently not available, are often too expensive to obtain or are incomplete, and what we must work with are their aggregated sample counterparts. For example, census data usually give only the frequency distribution of the number of individual units in each size class for each census year and report no information on the time path behavior of each individual. Alternatively, the individuals in the sample may vary from one time period to another and thus the only information we have is the proportion in each state at time t.

Faced with the restriction that in many cases only aggregated or total occurrence (proportion) sample data are available, the question arises as to whether or not it is possible to use the aggregate outcome data as a basis for estimating the transition probability matrix P which reflects or defines the behavior of the micro units. It is to this question that this book is directed and within this context, the purposes of this study are to
(1) develop estimators for the elements of the transition matrix P when only aggregated proportion sample data are available; and
(2) investigate, within the context of sampling experiments, the performance and distributional properties of the various estimators.

1.3. Plan of the book

Ch. 2 is devoted to a discussion of conventional and Bayesian estimators of the transition probabilities when the relevant micro sample data are available. Ch. 3 traces the early development of transition probability estimators which make use of aggregate time series data. In particular, a linear statistical model is specified, unrestricted and restricted least squares estimators are developed and their sampling properties discussed. In ch. 4, the model underlying the sampling experiments is specified and initially used as a basis for gauging the performance of the estimators developed in chs. 2 and 3.

In chs. 5–10, the weighted inequality restricted least squares, generalized least squares, minimum chi square, maximum likelihood, Bayesian and minimum absolute deviations estimators are developed and applied to the aggregate data generated from the sampling experiments and the actual data for two example problems. In chs. 11 and 12, the predictive ability of the various estimators are reviewed and their relative performance is evaluated. Ch. 13 is addressed to interpreting the results and commenting on areas for future research.

The generalized inverse method is discussed in appendix A and applied to the problem of developing a generalized least squares estimator for the transition probabilities. In appendix B, the linear probability model is discussed relative to the general problem of estimation when proportion data are used and several alternative estimators are specified. In appendix C, the estimation of variable transition probabilities is considered and an estimator to handle this problem is proposed. Finally, a computer routine for all of the estimators discussed in the book is given in appendix D.

CHAPTER 2

THE ESTIMATION OF TRANSITION
PROBABILITIES FROM MICRO DATA

Given the brief sketch of finite Markov chain theory in § 1.1, let us now consider the methods of statistical inference for this model when there are many observations in each of the initial states and the transition probabilities are *time constant*. Assuming that a sample of *micro data* exists and that there are repeated observations of the chain, we shall develop maximum likelihood and Bayesian estimators of the transition probabilities.

2.1. The micro maximum likelihood (ML) estimator[1]

Given the definitions and notation of § 1.1, let us assume that we have a sample of repeated observations on an ergodic Markov chain. Assume further that we are given $n_i(0)$ individuals in state i at time $t = 0$, and that the elements of an observation indicate the sequence of states the individuals are in at $t = 0, 1, ..., T$. Within the framework of § 1.1, let us specify the probability of an ordered sequence for a stationary Markov process as

$$\Pr(x_0, x_1, x_2, ..., x_T) = \Pr(x_0) \prod_t \Pr(x_t | x_{t-1}). \qquad (2.1.1)$$

If we let $n_{ij}(t)$ denote the number of individuals for which $x_{t-1} = s_i$ and $x_t = s_j$ and

$$n_{ij} = \sum_t n_{ij}(t), \qquad (2.1.2)$$

the probability of a given ordered set of sequences for the n individuals, within the context of (2.1.1), may be specified with $\propto$ denoting propor-

---

[1] The development of the ML estimator in this section follows closely the formulation given by Anderson and Goodman (1957).

tionality as
$$\Pr(x_0, x_1, ..., x_T | n) \propto \Pr(x_0) \prod_{i,j} p_{ij}^{n_{ij}}, \qquad (2.1.3)$$

and as shown by Anderson and Goodman (1957, p. 91), the n_{ij} form a set of sufficient statistics.

The distribution of the $n_{ij}(t)$ may be obtained by considering the $n_i(t-1) = \Sigma_j n_{ij}(t)$ observations on a multinomial distribution with probabilities p_{ij}. Within this context, the probability density function (PDF) of the $n_{ij}(t)$ is then

$$\Pr(n_{11}(t), n_{12}(t), ...|n(0)' \, p_{11}, ...) \qquad (2.1.4)$$

$$= \prod_t \left(\prod_i \left[\frac{n_i(t-1)!}{\prod_j n_{ij}(t)!} \prod_j p_{ij}^{n_{ij}(t)} \right] \right)$$

$$= \left[\prod_{t,i} \frac{n_i(t-1)!}{\prod_j n_{ij}(t)!} \right] \left[\prod_{i,j} p_{ij}^{\Sigma_t n_{ij}^{(t)}} \right]$$

$$= \left[\prod_{t,i} \frac{n_i(t-1)!}{\prod_j n_{ij}(t)!} \right] \left[\prod_{i,j} p_{ij}^{n_{ij}} \right],$$

where $n(0)' = [n_1(0), n_2(0), ..., n_r(0)]$ is a vector whose elements are the numbers in the states at time $t = 0$.

Given the observations $n_{ij}(t)$ for all i, j and t, we can obtain estimates of the stationary transition probabilities, the p_{ij}'s, by maximizing the likelihood function (2.1.4) with respect to the p_{ij}'s subject to the row sum conditions of (1.1.6). Proceeding in the usual fashion and expressing (2.1.3) in logarithmic form, we have the following Lagrangean function:

$$\log \Pr(x_0, x_1, ..., x_T | n) - \sum_i \lambda_i \left(\sum_j p_{ij} - 1 \right) \qquad (2.1.5)$$

or

$$\log \left(\prod_{ij} p_{ij}^{n_{ij}} \right) - \sum_i \lambda_i \left(\sum_j p_{ij} - 1 \right) + \text{constant}.$$

Since we are considering $n_i(0)$ as non-random, we have the following necessary conditions for (2.1.5) to have a maximum:

$$[\partial/\partial p_{ij}] \left[\sum_i \sum_j n_{ij} \log p_{ij} - \sum_i \lambda_i \left(\sum_j p_{ij} - 1 \right) \right] = n_{ij}/p_{ij} - \lambda_i = 0, \quad (2.1.6)$$

and

$$[\partial/\partial\lambda_j] \left[\sum_i \sum_j n_{ij} \log p_{ij} - \sum_i \lambda_i \left(\sum_j p_{ij} - 1\right)\right] = \sum_j p_{ij} - 1 = 0, \quad (2.1.7)$$

where a proportionality constant is omitted from (2.1.6) and (2.1.7). Rewriting the right-hand equality of (2.1.6) as $n_{ij} = \lambda_i p_{ij}$ and taking the sum of this expression with respect to j and making use of (2.1.7), we have

$$\lambda_i = \sum_j n_{ij}. \quad (2.1.8)$$

Substituting from (2.1.8) in (2.1.6) and solving for p_{ij}, we obtain the maximum likelihood estimator:

$$\dot{p}_{ij} = n_{ij}/\sum_j n_{ij} \geqq 0. \quad (2.1.9)$$

Since the n_{ij} are always non-negative, the ML estimator (2.1.9) also fulfills the non-negative constraint (1.1.5).

When viewed with the framework of asymptotic theory, as Kendall (1961, pp. 39–40, 42) has shown, the ML estimator (2.1.9) is consistent but it is not generally unbiased. However, Kendall (1961, p.42) shows that as the sample size increases, the bias tends to zero and Kendall (1961, pp. 43–44) and Anderson and Goodman (1957, p.95) show the estimates are asymptotically normally distributed. Anderson and Goodman (1957), using the results of Bartlett (1955) and Hoel (1954), show that the asymptotic properties for $T \to \infty$ and $n = 1$ are essentially the same as for the case we have considered. The same results also hold for the case where the elements of $n(0)$ are random variables.

Given ML estimates under the sampling scheme discussed above, Anderson and Goodman (1957, pp. 96–103) provide likelihood ratio tests and χ^2 tests of the type used in contingency tables for testing the following hypotheses: (1) the first order transition probabilities are time constant; (2) the transition probabilities are specified numbers and (3) the process is a given order Markov chain.

2.2. Bayesian analysis of the micro model

The preceding transition probability estimator has been derived on the basis of the sample information n_{ij} plus the requirements that the transition probabilities (1) cannot be negative, (2) cannot be larger than unity, and (3) the row sum of the probabilities must be unity.

There may be many cases in which an investigator has prior information about the structure of the individual elements of the transition probability matrix P. Such information may, for example, come from subject matter considerations and/or previous experimentation. If indeed such information is available, an investigator may wish to use it in addition to the sample information in making inferences about transition probabilities. As is well known, the Bayesian approach provides a convenient method for combining sample and prior information and it is to this approach that we now turn.

a. Bayes' theorem and Bayesian inference

In the Bayesian approach, it is assumed an investigator's information or uncertainty about a parameter vector, say θ, can be summarized in a prior probability density function pr (θ), with $\theta \in \Omega$, where Ω denotes an admissible region of the parameter space.[2] By the use of Bayes' theorem, this information can be combined with the sample density function pr $(y|\theta)$, $y \in R_y$, where y denotes a vector of sample observations and R_y, the sample space, to yield a posterior probability density function, pr $(\theta|y)$. This posterior PDF which contains all the sample and prior information can be used to make inferences about the parameters. Further, if an investigator has a loss function which reflects the losses due to incorrect estimation, it is generally possible to obtain an estimate, say $\hat{\theta}$, which minimizes the posterior expected loss. Under a wide range of conditions, $\hat{\theta}$ will also be the function of the sample observations which minimizes average risk. In this latter case, $\hat{\theta}$ is formally termed the Bayes' estimator relative to the given loss function and prior PDF employed. Bayes' estimators are known to be admissible and to constitute a complete class.[3]

Before analyzing the Markov probability model from the Bayesian point of view, let us first review Bayes' theorem. Let pr (y, θ) denote the joint probability density function of the observation vector y and the

---

[2] In denoting the prior PDF, Pr (θ), we do not explicitly show the parameters of the prior PDF which are assigned values by the investigator.

[3] For example, see Ferguson (1967, ch. 2).

parameter vector θ, with $y \in R_y$ and $\theta \in \Omega$. Then, from the usual operations with probability density functions, we have

$$pr\,(y, \theta) = pr\,(y|\theta)\,pr\,(\theta) \tag{2.2.1}$$
$$= pr\,(\theta|y)\,pr\,(y),$$

and

$$pr\,(\theta|y)\,pr\,(y) = pr\,(y|\theta)\,pr\,(\theta). \tag{2.2.2}$$

Thus, the posterior probability density function for the parameter vector θ, given the sample information y, is

$$pr\,(\theta|y) = [pr\,(\theta)\,pr\,(y|\theta)]/pr\,(y) \propto pr\,(\theta)\,pr\,(y|\theta), \qquad \text{for } \theta \in \Omega,$$
$$\tag{2.2.3}$$

where $\propto$ denotes proportionality, or

$$pr\,(\theta|y) \propto pr\,(\theta)\,l\,(\theta|y), \qquad \text{for } \theta \in \Omega, \tag{2.2.4}$$

where $pr\,(y|\theta)$, viewed as a function of θ, is the likelihood function, denoted $l\,(\theta|y)$. Eq. (2.2.4) is a statement of Bayes' Theorem, also known as the principle of inverse probability. Note that the posterior probability density function combines both prior and sample information, and it is this distribution which is employed in the Bayesian approach for making inferences.

b. The prior density function

In the transition probability model case, the parameters to be estimated have at least the following known properties:

(1) they cannot be negative;
(2) they cannot be larger than unity; and
(3) the sum of the probabilities of the exhaustive and mutually exclusive events must be unity.

In addition, from subject matter considerations and/or previous experiments, we often know the approximate values of certain parameters. To represent such information, we shall make use of a multivariate beta prior distribution. This is a rather rich distribution which we beleive can represent prior information adequately, in a number of circumstances.

Thus, following Martin (1967) and Mauldon (1961), let the prior probability density function (PDF) for the elements of the ith row of the transition probability matrix be a basic multivariate beta PDF:

$$f(p_{i1}, p_{i2}, ..., p_{ir}) = \frac{\Gamma(a_{i1} + a_{i2} + \cdots + a_{ir})}{\Gamma(a_{i1}) \, \Gamma(a_{i2}) \cdots \Gamma(a_{ir})} \cdot (p_{i1}^{a_{i1}-1} p_{i2}^{a_{i2}-1} \cdots p_{ir}^{a_{ir}-1}),$$

(2.2.5)

with $0 \le p_{ij} \le 1$, $j = 1, 2, ..., r$ and $\sum_j p_{ij} = 1$, where Γ denotes the gamma function and the a's are positive parameters to be assigned by the investigator to represent his prior information. The marginal PDF for any one of the p_{ij}'s is, of course, a standard univariate beta PDF.

There are r rows in the transition matrix P and thus there are r sets of transition probabilities for which a prior PDF must be formulated. If a priori the rows of P are assumed independently distributed, each in the multivariate beta form as shown in (2.2.5), then the joint PDF for all p_{ij} is

$$\prod_i f(p_{i1}, p_{i2}, ..., p_{ir}) = \prod_i \left(\frac{\Gamma\left(\sum_j a_{ij}\right)}{\prod_j \Gamma(a_{ij})} \prod_j p_{ij}^{a_{ij}-1} \right), \qquad (2.2.6)$$

for $0 \le p_{ij} \le 1$ and $i, j = 1, 2, ..., r$, which is a product of multivariate PDF's and may be called, following Martin (1967, p. 144), a special case of the matrix beta PDF.

To represent our prior information regarding the transition probabilities by (2.2.6), it is necessary to assign values to the a's, the parameters of the beta PDF. This can be done, given prior information about the means and variances of the p_{ij}'s, by noting that

$$E(p_{ij}) = a_{ij} / \sum_{j=1} a_{ij} = a_{ij}/a_i, \qquad \text{for } i, j = 1, 2, ..., r, \quad (2.2.7)$$

and

$$\text{Var}\,(p_{ij}) = \frac{a_{ij}\,(a_i - a_{ij})}{a_i^2\,(a_i + 1)} = \frac{E(p_{ij})\,(1 - E(p_{ij}))}{(a_i + 1)} \qquad \text{for } i, j = 1, 2, ..., r.$$

(2.2.8)

Thus, if the prior mean and variance, $E(p_{ij})$ and $\text{Var}\,(p_{ij})$, are assigned, a_i may be determined by (2.2.8) and a_{ij} may be obtained from (2.2.7).

As shown by Martin (1967, p. 140), in fixing $E(p_{ij})$ and $E(p_{ik})$ and getting a_i, the a priori covariances will be negative.[4]

c. The posterior probability density function

Given the likelihood for the transition probabilities p_{ij} (2.1.3) and the matrix beta PDF as our prior PDF, we may proceed to apply Bayes' theorem. Given the micro sample data, the n_{ij}'s, the likelihood function is

$$l(P|n) \propto \prod_{i,j} p_{ij}^{n_{ij}} \tag{2.2.9}$$

$$\propto \prod_{i=1}^{r} \prod_{j=1}^{r-1} p_{ij}^{n_{ij}} \left(1 - \sum_{j=1}^{r-1} p_{ij}\right)^{n_{ir}},$$

with the p_{ij}'s constrained by $0 \le p_{ij} \le 1$ and $\Sigma_j p_{ij} = 1$ and where n denotes the vector with elements n_{ij} $(i, j = 1, 2, ..., r)$. Then using Bayes' theorem with the PDF (2.2.6), the posterior PDF is

$$\Pr(P|n) \propto \Pr(P)\, l(P|n) \propto \prod_{i,j} p_{ij}^{n_{ij}+a_{ij}-1} \tag{2.2.10}$$

$$\propto \prod_{i=1}^{r} \prod_{j=1}^{r-1} p_{ij}^{n_{ij}+a_{ij}-1}\left(1 - \sum_{j=1}^{r-1} p_{ij}\right)^{n_{ir}+a_{ir}-1},$$

for $0 \le p_{ij} \le 1$ and $\Sigma_j p_{ij} = 1$ $(i = 1, 2, ..., r)$.

Thus, the posterior PDF for the elements of P when the proportional constant is inserted is again a PDF in the form of a product of multivariate beta PDF's with parameters $n_{ij} + a_{ij}$:

$$\Pr(P|n) = \prod_{i=1}^{r} \left(\frac{\Gamma\left(\sum_{j=1}^{r} n_{ij} + a_{ij}\right)}{\prod_{j}^{r} \Gamma(n_{ij} + a_{ij})} \prod_{j}^{r} p_{ij}^{n_{ij}+a_{ij}-1} \right), \tag{2.2.11}$$

$$= k \prod_{i=1}^{r} \prod_{j=1}^{r-1} p_{ij}^{n_{ij}+a_{ij}-1} \left(1 - \sum_{j=1}^{r-1} p_{ij}\right)^{n_{ir}+a_{ir}-1}.$$

---

[4] See Martin (1967, p. 146) for a discussion of an 'extended natural conjugate' prior distribution which admits non-zero correlation between the rows of P. As Martin notes, relaxing the independence assumption results in complicated formulas for the moments, etc.

As Martin (1967, p. 144) shows, the marginal posterior PDF of a sub-matrix of P is also in matrix beta form. Further, the marginal posterior PDF for a single transition probability is in the standard beta form with mean and variance given by

$$E(p_{ij}|n) = \frac{n_{ij} + a_{ij}}{\sum\limits_{j=1}^{r} (n_{ij} + a_{ij})} = \frac{n_{ij} + a_{ij}}{c_i}, \qquad (2.2.7a)$$

and

$$\text{Var}(p_{ij}|n) = \frac{(n_{ij} + a_{ij})(c_i - n_{ij} - a_{ij})}{c_i^2 (c_i + 1)} \qquad (2.2.8a)$$

$$= \frac{E(p_{ij}|n)(1 - E(p_{ij}|n))}{c_i + 1}.$$

d. Bayesian estimation

The analysis above yields the complete posterior PDF for the transition probabilities, a PDF which incorporates sample and prior information. The posterior PDF can be employed to make inferences about the transition probabilities. With respect to point estimation, it is well known that the mean of the posterior PDF is the Bayesian estimator given that our loss function is quadratic while the median of the posterior PDF is optimal for an absolute error loss function. Further, it is of interest to point out that the mode of the posterior PDF is associated with the following values for transition probabilities:

$$\ddot{p}_{ij} = \frac{n_{ij} + a_{ij} - 1}{\sum\limits_{j} n_{ij} + \sum\limits_{j} a_{ij} - r}, \qquad \text{for } i = 1, 2, ..., r \text{ and } j = 1, 2, ..., r-1.$$
$$(2.2.12)$$

This result can be obtained by maximizing the logarithm of (2.2.11) with respect to the p_{ij}'s. The values of the remaining r transition probabilities p_{ir}, $i = 1, 2, ..., r$, can be obtained from $\Sigma_{j=1}^{r} p_{ij} = 1$, $i = 1$, $2, ..., r$. Blackwell and Girschick (1954, p. 305) provide a loss structure for which the modal value of a posterior PDF is an optimal point estimate.

CHAPTER 3

THE ESTIMATION OF TRANSITION
PROBABILITIES FROM MACRO DATA

In the preceding chapter, maximum likelihood and Bayesian estimators were developed under the assumption that time ordered *micro* observations reflecting repeated observations of the chain were available. This chapter and the transition probability estimators developed in the following chapters proceed under the assumption that the individual ime traces, $n_{ij}(t)$, relative to the sequence of states are unavailable and only the sample aggregate proportions relating to the number of individuals in each state for each time period t are known.

3.1. A relation involving the macro data

If the $n_{ij}(t)$ sample observations are not available and only the aggregate outcome data, $n_j(t)$, which are equal to $\Sigma_i n_{ij}(t)$, are available, then one way to make use of the observed proportion data in estimating the transition probabilities is, using the notation of § 1.1, to make use of the argument of condition probability in the following way:

$$\Pr(x_{t-1} = s_i, x_t = s_j) = \Pr(x_{t-1} = s_i) \Pr(x_t = s_j | x_{t-1} = s_i). \quad (3.1.1)$$

Given (3.1.1) and using the generalized addition law of the probability, then

$$\Pr(x_t = s_j) = \sum_i \Pr(x_{t-1} = s_i) \Pr(x_t = s_j | x_{t-1} = s_i), \quad (3.1.2)$$

or

$$q_j(t) = \sum_i q_i(t-1) p_{ij}, \quad (3.1.3)$$

where $q_j(t)$ and $q_i(t-1)$ represent the unconditional probabilities $\Pr(x_t = s_j)$ and $\Pr(x_{t-1} = s_i)$ respectively. If the unconditional probabilities $q_j(t)$ and $q_i(t-1)$ in eq. (3.1.3) are replaced by the actual observed proportions $y_j(t)$ and $y_i(t-1)$, then there will be no set of

31

transition probabilities that will satisfy this relation with probability one. Thus, if errors are admitted in eq. (3.1.3) to account for the difference between the actual and estimated occurrence of $y_j(t)$, based on the right-hand side of eq. (3.1.3), then the sample observations may be assumed to be generated by the following stochastic relation:

$$y_j(t) = \sum_i y_i(t-1)p_{ij} + u_j(t). \tag{3.1.4}$$

Miller (1952) proposed using this stochastic relation as a basis for specifying a linear statistical model for estimating the transition probabilities. It is to this proposition that we now turn.

3.2. The unrestricted least squares transition probability estimator

In developing Miller's (1952) approach for estimating the transition probabilities in (3.1.4) from sample proportion data, let us rewrite the stochastic relation given in (3.1.4) in matrix form as

$$y_j = X_j p_j + u_j, \tag{3.2.1}$$

where $y_j = \{y_j(t)\}$ is a $(T \times 1)$ vector of sample proportions, $p'_j = (p_{1j}, p_{2j}, ..., p_{rj})$ is a $(r \times 1)$ vector of unknown transition parameters to be estimated, u_j is a $(T \times 1)$ vector of random disturbances and X_j is the following $(T \times r)$ matrix:

$$X_j = \begin{bmatrix} y_1(0) & y_2(0) & \cdots & y_r(0) \\ \vdots & \vdots & & \vdots \\ y_1(t-1) & y_2(t-1) & \cdots & y_r(t-1) \\ \vdots & \vdots & & \vdots \\ y_1(T-1) & y_2(T-1) & \cdots & y_r(T-1) \end{bmatrix}. \tag{3.2.2}$$

It is assumed that the matrix X_j has rank r. We make the following assumptions about the random disturbance vector u_j in (3.2.1):

$$E(u_j) = 0, \tag{3.2.3}$$

and

$$E(u_j u'_j) = \sigma_j \omega_{jj}, \tag{3.2.4}$$

where ω_{jj} is a $(T \times T)$ positive definite diagonal matrix.[1]

---

[1] For the basis underlying the residual specification, see Madansky (1959).

The set of equations of which (3.1.4) or (3.2.1) is a part, may then be written as

$$
\begin{bmatrix} y_1 \\ y_2 \\ \vdots \\ y_r \end{bmatrix} = \begin{bmatrix} X_1 & 0 & \cdots & 0 \\ 0 & X_2 & \cdots & 0 \\ \vdots & \vdots & \ddots & \vdots \\ 0 & 0 & \cdots & X_r \end{bmatrix} \begin{bmatrix} p_1 \\ p_2 \\ \vdots \\ p_r \end{bmatrix} + \begin{bmatrix} u_1 \\ u_2 \\ \vdots \\ u_r \end{bmatrix} \tag{3.2.5}
$$

or

$$
y = Xp + u, \tag{3.2.6}
$$

with $y' = (y_1', y_2', ..., y_r'), p' = (p_1', p_2', ..., p_r'), u' = (u_1', u_2', ..., u_r')$ and X is the block diagonal matrix on the right-hand side of (3.2.5) with $X_1 = X_2 = \cdots = X_r$,

$$
E(u) = 0, \tag{3.2.7}
$$

and

$$
E(uu') = \Sigma, \tag{3.2.8}
$$

where Σ is a $((Tr) \times (Tr))$ non-diagonal, singular matrix.[2]

Given the multivariate linear statistical model (3.2.5) or (3.2.6),

$$
y = Xp + u, \tag{3.2.6}
$$

and assuming T strictly greater than r, Miller (1952) suggested the use of the conventional least squares estimator as a basis for obtaining estimates of the transitional probabilities. That is, he viewed the problem as one of finding the estimate $\tilde{p}$ which minimizes the positive definite quadratic form

$$
\phi = u'u = (y - Xp)'(y - Xp). \tag{3.2.9}
$$

Solving this extremum problem in the conventional way yields the minimizing solution

$$
\tilde{p} = (X'X)^{-1}X'y, \tag{3.2.10}
$$

provided that $X'X$ is non-singular. Since $X_j, j = 1, 2, ..., r$, has been assumed to be of rank r, $X'X$, a block diagonal matrix with matrices $X_j'X_j$ on the main diagonal, will be non-singular. Since the matrix $X'X$ is positive definite (and also symmetric), both the necessary and sufficient conditions for $\tilde{p}$ to minimize (3.2.9) are fulfilled.

---

[2] Singularity of Σ results since $\Sigma_j y_j - \Sigma_j X_j p_j = \eta_T - X_j \eta_r = \eta_T - \eta_T = 0$ means the rows of Σ are linearly dependent where η_T and η_r are vectors of order T and r with all elements unity.

Although the set of relations (3.2.5) are 'disturbance related', since $X_1 = X_2 = \cdots = X_r$, the j equations may be estimated separately or as a set (Zellner 1962) with the same results.

Thus, the unrestricted 'conventional' least squares estimator for p_j, a subvector of $\tilde{p}$, is, from (3.2.10),

$$\tilde{p}_j = (X'_j X_j)^{-1} X'_j y_j, \qquad j = 1, 2, ..., r. \qquad (3.2.11)$$

The question that now arises is whether or not these conventional least squares estimates of the transition probabilities, p_{ij}, satisfy the following non-negativity and row sum conditions:

$$0 \le p_{ij} \le 1 \qquad (3.2.12)$$

and

$$\sum_j p_{ij} = 1, \qquad \text{for all } i. \qquad (3.2.13)$$

a. The row sum condition

To specify the problem so as to take account of the row sum condition (3.2.13), we reformulate the problem as one of minimizing the sum of the squared errors,

$$\phi = u'u = (y - Xp)'(y - Xp), \qquad (3.2.9)$$

subject to

$$Gp = \eta_r \qquad (3.2.14)$$

or

$$\sum_j p_{ij} = 1, \qquad \text{for all } i, \qquad (3.2.13)$$

where G is an $(r \times r^2)$ known coefficient matrix $[I_1, I_2, ..., I_r]$ with each I_i an $(r \times r)$ identity matrix and η_r is an $(r \times 1)$ column vector with all entries equal to one.[3]

Without imposing the non-negativity restriction

$$p \ge 0, \qquad (3.2.15)$$

we may take into account the row sum condition (3.2.14) and solve the partially restricted problem by forming the following Lagrangean:

$$\phi(p, \lambda) = (y - Xp)'(y - Xp) - 2\lambda'(Gp - \eta_r), \qquad (3.2.16)$$

---

[3] For discussion of least squares estimation under exact linear restraints, see Theil (1963) and Goldberger (1964, pp. 256–257). For a more complete development of the theorems in this and the following section, see Takayama, Judge and Lee (1969).

where λ is an $(r \times 1)$ vector of Lagrange multipliers and finding the saddle point (p^*, λ^*) of $\phi(p, \lambda)$, which is

$$p^* = (X'X)^{-1} X'y + (X'X)^{-1} G'(G(X'X)^{-1}G')^{-1} (\eta_r - G(X'X)^{-1} X'y), \tag{3.2.17}$$

and

$$\lambda^* = (G(X'X)^{-1} X'G')^{-1} (\eta_r - G(X'X)^{-1}X'y). \tag{3.2.18}$$

Since from (3.2.10) the unrestricted least squares estimates are

$$\tilde{p} = (X'X)^{-1} X'y, \tag{3.2.10}$$

then (3.2.17) and (3.2.18) may be written as

$$p^* = \tilde{p} + (X'X)^{-1} G'(G(X'X)^{-1}G')^{-1}(\eta_r - G\tilde{p}), \tag{3.2.19}$$

and

$$\lambda^* = (G(X'X)^{-1} X'G')^{-1} (\eta_r - G\tilde{p}). \tag{3.2.20}$$

In order to show that the row sum restriction (3.2.14) is automatically fulfilled, it is sufficient to prove any one of the following equivalent statements:

$$p^* = \tilde{p}, \tag{3.2.21}$$

$$\lambda^* = 0, \tag{3.2.22}$$

and

$$G\tilde{p} - \eta_r = 0. \tag{3.2.23}$$

In the implication ordering of these statements, (3.2.23) implies (3.2.22), (3.2.22) implies (3.2.21) and finally (3.2.21) implies (3.2.23). Thus, in order to conclude that $p^* = \tilde{p}$, it is sufficient to prove (3.2.23) or (3.2.22). We prove (3.2.23) first.

Theorem 1: $G\tilde{p} = \eta_r$.
Proof: Since $G = (I I \cdots I); (r \times r^2)$, we have

$$G\tilde{p} = [I I \cdots I](X'X)^{-1} X'y \tag{3.2.24}$$

$$= [I I \cdots I] \begin{bmatrix} (X_1'X_1)^{-1} & & & \\ & (X_2'X_2)^{-1} & & \\ & & \ddots & \\ & & & (X_r'X_r)^{-1} \end{bmatrix} \begin{bmatrix} X_1' & & & \\ & X_2' & & \\ & & \ddots & \\ & & & X_r' \end{bmatrix} \begin{bmatrix} y_1 \\ y_2 \\ \vdots \\ y_r \end{bmatrix} \tag{3.2.25}$$

$$= ((X_1'X_1)^{-1}X_1', (X_2'X_2)^{-1}X_2', \ldots, (X_r'X_r)^{-1}X_r') \begin{bmatrix} y_1 \\ y_2 \\ \vdots \\ y_r \end{bmatrix} \qquad (3.2.26)$$

$$= (X_1'X_1)^{-1}X_1' (y_1 + y_2 + \cdots + y_r) \qquad (3.2.27)$$

$$= (X'X)^{-1}X'\eta_T, \qquad (3.2.28)$$

and since $\eta_T = X_1\eta_r$ the above equality is reduced to

$$G\tilde{p} = (X_1'X_1)^{-1}X_1'X_1\eta_r \qquad (3.2.29)$$

$$= \eta_r.$$

This result shows that $G\tilde{p} = \eta_r$ automatically holds and thus restates the Goodman (1953) conclusion for this model.[4]

b. The $0 \leq p_{ij} \leq 1$ condition

As noted earlier, in order to qualify as a transition probability estimate, both conditions (3.2.12) and (3.2.13) must be fulfilled. We have shown, following Goodman (1953), that the row sum condition (3.2.13) is automatically met. However, the non-negativity[5] condition (3.2.12) may be violated by the unrestricted least squares estimator.

To show this, let us rewrite the unrestricted least squares estimator (3.2.10) as the following equation system:

$$(X'X)\tilde{p} = X'y. \qquad (3.2.10a)$$

---

[4] By using some of the properties of the matrices developed in the proof of Theorem 1, it is easy to prove directly, as Telser (1963) implied, the following:

Theorem 2: $\lambda^* = (G (X'X)^{-1} G')^{-1} (\eta_r - G\tilde{p}) = 0$
Proof: Since $G (X'X)^{-1} G' = r (X_1'X_1)^{-1}$, $(G (X'X)^{-1} G')^{-1} = (1/r) X_1'X_1$. Therefore, we get directly:

$\lambda^* = (G (X'X)^{-1} G')^{-1} (\eta_r - G\tilde{p}) = (1/r) (X_1'X_1) (\eta_r - (X_1'X_1)^{-1} X_1'\eta_r)$
$= (1/r) (X_1'X_1\eta_r - X_1'\eta_T) = (1/r) (X_1'\eta_T - X_1'\eta_T) = (1/r) (\eta_r - \eta_r) = 0.$

[5] Correspondingly, of course, this also implies that the less than or equal to one condition may be violated.

In this expression,

$$X'X = \begin{bmatrix} X_1'X_1 & & & \\ & X_2'X_2 & & \\ & & \ddots & \\ & & & X_r'X_r \end{bmatrix} \qquad (3.2.30)$$

is an $(r^2 \times r^2)$ matrix and $X_i'X_i$ is an $(r \times r)$ matrix of *non-negative* elements and

$$X'y = \begin{bmatrix} X_1 y_1 \\ X_2 y_2 \\ \vdots \\ X_r y_r \end{bmatrix}, \qquad (3.2.31)$$

an $(r^2 \times 1)$ *non-negative* vector. Thus, we are working with a system of equations with a non-negative right-hand side vector and a *non-negative* matrix $X'X$.

As a framework for considering the non-negative solvability of (3.2.10a), let

$$(X'X)\tilde{p} = d'(I - A)\,d\tilde{p}, \qquad (3.2.10b)$$

where d is a non-singular positive diagonal matrix. Therefore, (3.2.10a) may be rewritten as

$$(I - A)\,d\tilde{p} = [d]^{-1} X'y \geq 0 \qquad (3.2.10c)$$

or

$$(I - A)\,\tilde{w} = c \geq 0. \qquad (3.2.10d)$$

If A is a non-negative block diagonal matrix with $0 \leq a_{ij} \leq 1$ where $\Sigma_{j=1}^r a_{ij} < 1$ and $\Sigma_{i=1}^r a_{ij} < 1$ for all i and j, then the following theorem may be used:

Theorem 3: $(I - A)\,\tilde{w} = B\tilde{w} = c \geq 0$ has a non-negative solution for $\tilde{w}$ if and only if the following conditions hold:

$$b_{11} > 0; \begin{vmatrix} b_{11} & b_{12} \\ b_{21} & b_{22} \end{vmatrix} > 0, \ldots, \begin{vmatrix} b_{11} & \cdots & b_{1n} \\ \vdots & & \vdots \\ b_{n1} & \cdots & b_{nn} \end{vmatrix} > 0. \qquad (3.2.32)$$

This so-called Hawkins–Simons condition (1949) is equivalent to saying that for $\tilde{w}$ to be non-negative, the vector c must be contained in the convex cone spanned by the column vectors of the B or $(I - A)$ matrix. However, since $X'X$ is a positive definite matrix with non-negative elements, condition (3.2.32) is satisfied but the corresponding non-negative requirements on the A matrix in (3.2.10d) are violated. Therefore, in general, the non-negativity of the unrestricted least squares estimate, $\tilde{p}$, *is not* automatically fulfilled.

c. *Some properties of the unrestricted estimator* $\tilde{p}$

As regards to the properties of the unrestricted estimator $\tilde{p}$, Madansky (1959) has shown that u_j is uncorrelated with X_j and thus the unrestricted least squares estimator of the transition probabilities converges in probability to p as $T \rightarrow \infty$. Furthermore, for a fixed T, Madansky (1959) has shown that the conventional least squares estimator $\tilde{p}$ has the probability limit p as $n \rightarrow \infty$ and thus is consistent.[6]

---

[6] To show some of the properties of the sample data and that the unrestricted least squares estimator $\tilde{p}$ is not unbiased, let $q_i(t)$ be the probability of an individual being in state i at time t. Assume the population is large (or otherwise sampling is with replacement) so that the probability of obtaining an individual in state i at any drawing is not affected by other drawings and is a constant, $q_i(t)$. Let $y_i(t)$ be the proportion observed in time t for state i of a multinomial population based on a sample of size $N(t)$. Then $N(t) y_i(t)$ represents the individuals in state i at time t and $N(t)(1 - y_i(t))$ represents those individuals not in state i at time t. The expected value of $y_i(t)$ over all possible samples is $q_i(t)$, and thus our proportion estimate is unbiased. If we then let $\overline{X}$ and $\overline{y}$ be the appropriate data of proportions for all time periods from the population, and let X_s and y_s reflect the same data from a sample of size $N(t)$, then the least squares estimates from the population and sample are:

$$p = (\overline{X}'\overline{X})^{-1} \overline{X}'\overline{y}, \tag{F.1}$$

and

$$\tilde{p}_s = (X_s'X_s)^{-1} X_s'y_s. \tag{F.2}$$

Since

$$E(y_i(t)) = q_i(t), \tag{F.3}$$

therefore

$$E(X_s) = \overline{X} \quad \text{and} \quad E(y_s) = \overline{y}. \tag{F.4}$$

However, $\tilde{p}_s$ is *not an unbiased* estimator of p since

$$E(\tilde{p}_s) = E\left[(X_s'X_s)^{-1} X_s'y_s\right] \neq (EX_s'EX_s)^{-1} EX_s'Ey_s = (\overline{X}'\overline{X})^{-1} \overline{X}'\overline{y} = p, \tag{F.5}$$

due to the dependencies of X_s, X_s' and y_s. Furthermore, since

$$\text{plim } y_i(t) = \text{plim } n_i(t)/N(t) = q_i(t), \tag{F.6}$$

$y_i(t)$ is a consistent estimator of $q_i(t)$. This implies that

$$\text{plim } X_s = \overline{X} \quad \text{and} \quad \text{plim } y_s = \overline{y}. \tag{F.7}$$

However, since this problem involves proportions, heteroscedasticity is present. That is, the covariance matrix for the disturbances, given in (3.2.3), is not of a form of a positive scalar times an identity matrix. Since the unweighted least squares estimator (3.2.10) does not take account of heteroscedasticity, this estimator is not efficient.[7]

3.3. The restricted least squares transition probability estimator

Theorems 1 and 3 mean that although the unrestricted least squares estimates automatically fulfill condition (3.2.13), the non-negativity condition (3.2.15) may not be satisfied and un-acceptable estimates of the transition probabilities may result. The prospect of non-negative transition probability estimates was viewed by Goodman (1953, p. 247) as follows:

> ... The problem is that of estimating $(a \times a)$ parameters which are subject to linear constraints. We shall be interested in minimizing $\Sigma c_i c_i$ in order to obtain the least squares estimate. In other words, we wish to estimate a parameter which is a point in an $a(a-1)$ dimensional space. Since $t_{ij} \geq 0$ the parameter will lie in a subset of this space. If we also wish our estimate $\hat{T}$ to lie in the same subset the method of estimation is still straight forward but tedious. We first obtain the least squares estimates $\hat{T}$. If this estimate lies in the subset then $\hat{T}$ is used to estimate T. If $\hat{T}$ is not included in the subset then the appropriate estimate will lie on the boundary of the subset. We then use the estimate on the subset which minimizes $\Sigma c_i c_i$ ...

(In our notation, $a = r$, $t_{ij} = p_{ij}$, $T = P$ and $\Sigma c_i c_i = u_j' u_j$).

Thus, Goodman (1953) recognized that the conventional least squares estimator may violate the condition $0 \leq p_{ij} \leq 1$, and suggested that in this case the appropriate estimate will lie on the boundary of the restricted parameter subset and concluded that the estimate on the boundary of the subset which minimizes the quadratic form (3.2.9) should be used. Telser (1963), who made use of the procedure proposed

---

[7] For a discussion of the heteroscedasticity problem as it relates to the estimation of transition probabilities, see Madansky (1959), and for estimating methods to use with proportion data, see Zellner and Lee (1965).

by Miller (1952) and Goodman (1953), suggests a subjective adjustment procedure to correct the transition probability estimates falling outside of the zero to one interval. Based on Telser's work (1963), Lee, Judge and Takayama (1965) and Theil and Rey (1966) proposed an inequality restricted estimator which satisfied Goodman's conditions. They chose estimates which minimize

$$u'u = (y - Xp)'(y - Xp), \tag{3.3.1}$$

subject to constraints

$$Gp = \eta_r, \tag{3.3.2}$$

$$p \geq 0. \tag{3.3.3}$$

Since (3.3.1) appears as a quadratic form in p and the restrictions are linear, this problem is a typical quadratic programming problem. Following Judge and Takayama (1966), by making use of the Kuhn–Tucker (1951) equivalence theorem for non-linear programming and the duality theorem of Dorn (1960) for quadratic programming, we may reduce the problem to the following linear programming specification: Find $\tilde{p}^c$ that maximizes

$$(X'y - X'X\tilde{p}^c)'p, \tag{3.3.4}$$

subject to

$$Gp \leq \eta_r, \tag{3.3.5}$$

$$-Gp \leq -\eta_r, \tag{3.3.6}$$

and

$$p \geq 0, \tag{3.3.7}$$

where $\tilde{p}^c$ is the optimal restricted least squares estimator.

By virtue of the duality theorem of linear programming (Dantzig and Orden 1953), the corresponding dual linear programming is to minimize

$$[\lambda_1' \lambda_2'] \begin{bmatrix} \eta_r \\ -\eta_r \end{bmatrix}, \tag{3.3.8}$$

subject to

$$[G' -G'] \begin{bmatrix} \lambda_1 \\ \lambda_2 \end{bmatrix} \geq X'y - X'X\tilde{p}^c, \tag{3.3.9}$$

and

$$\lambda_1, \lambda_2 \geq 0, \tag{3.3.10}$$

where λ_1 and λ_2 are $(r \times 1)$ vectors of dual variables.

In order to develop a solution algorithm, let us remove the $\tilde{p}^c$ restrictions on p in the primal and dual specifications and define the following primal–dual programming formulation: To maximize

$$(X'y - X'Xp)'p - \lambda_1'\eta_r + \lambda_2'\eta_r = -\lambda_1'\alpha_1 - \lambda_2'\alpha_2 - \beta'p \leq 0, \quad (3.3.11)$$

subject to

$$Gp = \eta_r, \quad (3.3.12)$$

$$G'\lambda_1 - G'\lambda_2 + (X'X)p - \beta = X'y, \quad (3.3.13)$$

and

$$p, \lambda_1, \lambda_2, \alpha_1, \alpha_2, \beta \geq 0, \quad (3.3.14)$$

where α_1, α_2 and β are the vectors of slack variables to primal and dual respectively.

The above problem can be readily solved by the use of the standard simplex version of the quadratic programming algorithm developed by Wolfe (1959). The characteristics of the formulation and algorithm are reflected in the tableau given in table 3.1.

TABLE 3.1

Quadratic programming simplex tableau for the classical restricted least squares estimator

B_0	λ_1	λ_2	p	α_1	α_2	β
η_r			G	I		
$-\eta_r$			$-G$		I	
$X'y$	G'	$-G'$	$X'X$			$-I$

a. Sampling properties of the restricted estimator

The sampling properties of restricted estimators, when the restrictions occur in equality form as in (3.3.2), have been investigated by Theil (1961). When the restrictions occur in inequality form as (1.1.5) or (3.2.15) and quadratic programming procedures are used to derive the estimates, the sampling properties of a restricted least squares estimator for a single regressor have been investigated by Zellner (1961). Within this context, Zellner considered the problem for the case of one fixed variable when the u vector is assumed to have a multivariate normal distribution with a mean of zero and a variance of $\sigma^2 I$. Thus, as is well known, the *unconstrained* least squares estimator for β is distributed normally with a mean of β and a variance of $\sigma^2 (X'X)^{-1}$.

In contrast to the sampling distribution of the unrestricted estimator, as noted by Zellner (1961), inequality restricted estimators have distributions which are of the truncated normal form[8] (partly continuous and partly discrete). Under the inequality restraint specification as reflected by (1.1.5) or (3.2.15) for a single fixed variable the continuous part of the distribution has the truncated normal form. For simple regression problems, we can derive and investigate the moments of the least squares estimator subject to inequality constraints. However, when more than one independent variable is involved, it is difficult to evaluate the moments and obtain the sampling properties of the restricted estimators using analytical techniques. The basis of this difficulty is that the restrictions that are binding are not constant from sample to sample and must be considered as a vector of random variables. Thus, the inequality restricted least squares estimator for the standard linear statistical model is the sum of the products of the random variables (effective restriction set and y) instead of a linear combination of random variables as in the case of unrestricted least squares or equality restricted least squares estimators. Some properties of the inequality restricted least squares estimator are discussed by Hartley (1963) and the exact distribution has been derived by Hocking (1963). However, as noted by Hocking, the complexity of the exact distribution limits its use, since the determination of the moments is, in general, not a simple problem.

---

[8] For a discussion of distributions of this form, see Cramer (1946).

CHAPTER 4

THE SAMPLING EXPERIMENT
AND SOME INITIAL RESULTS

Given the difficulty of determining the sampling properties of the inequality restricted least squares transition probability estimators and the fact that for many of the other estimators for transition probabilities, only asymptotic results are available, this chapter is concerned with specifying a probability model that may be used to generate data that will form a basis for evaluating the finite sample performance of alternative estimators via Monte Carlo experiments. A basis for gauging the performance of alternative estimators is discussed and the sampling results for the micro and macro estimators, presented in chs. 2 and 3, are presented.

In interpreting the results given in this and succeeding chapters, we want to be quite clear relative to the limitations placed on the results gained through the sampling experiment approach. Thornber (1967) and others have pointed out that conclusions regarding the properties of estimators may vary depending on the parameter value(s) chosen for the experiment. The implication of this proposition is that we need to develop the risk function over the entire parameter space rather than at just one or more isolated points to compare the properties of estimators. Unfortunately, the results to be given in this and the following chapters follow from sampling only one point in the parameter space and thus our basis for generalization is severely restricted. However, the results do point up differences in the properties of estimators at the point selected and provide information on the speed of convergence to large sample results. Furthermore, it is well known that the Bayesian estimator relative to a given loss function and prior PDF is the one, provided it exists, that minimizes average risk.

In addition, the criterion of performance in 'repeated samples' has been used as a basis for gauging the effectiveness of various estimators. There is a question, of course, whether or not this is the most meaningful criterion to employ since, for example, in the social science area, one does not usually have the option of this repeated sampling process.

4.1. The simulated probability model and its characteristics

The simple first-order stationary Markov process used as a basis for the sampling study employs the following transition probability matrix:

$$P = [p_{ij}] = \begin{matrix} & \begin{matrix} S_1 & S_2 & S_3 & S_4 \end{matrix} \\ \begin{matrix} S_1 \\ S_2 \\ S_3 \\ S_4 \end{matrix} & \begin{bmatrix} 0.60 & 0.40 & 0.00 & 0.00 \\ 0.10 & 0.50 & 0.40 & 0.00 \\ 0.00 & 0.10 & 0.70 & 0.20 \\ 0.00 & 0.00 & 0.10 & 0.90 \end{bmatrix} \end{matrix}. \tag{4.1.1}$$

The behavior pattern reflected by this transition matrix indicates (1) a strong tendency of individuals to remain within a given state from one time period to the next, and (2) the most probable outcome, excluding remaining in the same state, is that individuals either move up or down one state at a time. This type of behavior appears to be consistent with that observed for many kinds of economic choice units. The regular Markov matrix (4.1.1) has as its equilibrium or stationary vector,

$$\pi = (0.0189 \quad 0.0755 \quad 0.3020 \quad 0.6036). \tag{4.1.2}$$

That is, as the time period t increases, each row of P^t approaches (4.1.2), or in general,

$$P^t \to \eta_r \pi \qquad \text{as } t \to \infty, \tag{4.1.3}$$

where η_r is a column vector with r elements all unity. For the transition matrix (4.1.1), if the process is started in state 1, it approaches the stationary vector (4.1.2) in approximately 40 time periods.

Given the transition matrix (4.1.1), we need an initial probability vector to define a Markov process. If we let $y(0)$ be the $(1 \times r)$ initial vector (proportion data) and $y(t)$ be the $(1 \times r)$ probability vector (or aggregate data) at time t, then from the definition of a Markov process,

$$y(t) = y(0) P^t. \tag{4.1.4}$$

Obviously, different starting vectors $y(0)$ generate different patterns of aggregate data $y(t)$. In the model (4.1.1), there are four possible starting states and the patterns of changes in the aggregate data for the different starting states are shown in fig. 4.1. As reflected by fig. 4.1, when starting in state 1, the observations of the probability vector $y_j(t)$ exhibit their greatest variability in the initial time periods of the process and then approach the equilibrium vector as a geometrically decreasing sequence.

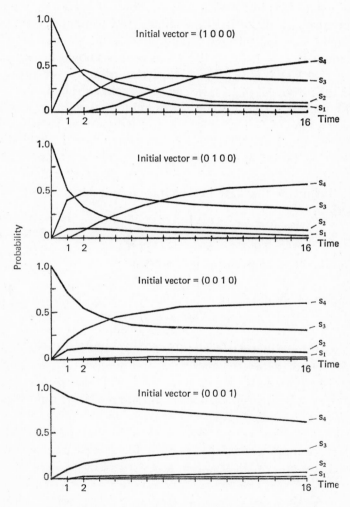

Fig. 4.1. Alternative initial vectors and time paths.

4.2. The procedure of simulation

The probability model (4.1.1) was employed to investigate the properties of the restricted least squares estimators for transition probabilities.
The sampling experiments were performed as follows: the sampling
process for the stationary Markov process reflected by (4.1.1) was
started in state 1 and by making use of tables of random permutations
of the numbers from 1 to 1000 and an IBM 7094 computer, one thousand
'individuals' were led through a random walk of 24 steps. For example,
if individual k ($k = 1, 2, ..., 1000$) at step t is known to be in state i,
where for (4.1.1) $i = 1, 2, 3$ or 4, then according to the probabilities
given in the ith row of (4.1.1), we can specify that if the tth random digit
in the ith table is $< 10p_{i1}$, then the kth individual will step to state 1, if
the tth random digit is $\geq 10p_{i1}$ but $< 10(p_{i1} + p_{i2})$, the individual
will step to state 2, etc. The process is repeated for all 1000 individuals
in the ith time period and then the process is repeated for the $(t + 1)$st
time period, etc. Thus, by this process, a (1000×24) data matrix is
generated showing the state in which each individual is found in each
time period.

4.3. The simulated population and sampling

The resulting aggregate proportion data for the 1000 individuals (micro
units) for twenty time periods are given in table 4.1.
    From the population of one thousand individuals, sampling is per-
formed at three different levels: 25, 50 and 100 individuals. At each
level, 50 sets of data are obtained. Time periods 1 through 16 or 2
through 14 are used as the time span basis of the samples of data. Fifty
estimates are obtained in each level (or sample size); their means and
dispersions are studied and the comparisons are made among three
levels. Since the time periods and the 50 sets of data for each level are
fixed, we are interested in the effect of the sample size or number of
individuals[1] on the estimates.

---

[1] We emphasize here that the sample size refers to the number of individuals and
not to the number of periods or sets of data.

TABLE 4.1

The aggregate data from a simulation study of 1000 individuals

Time period	S_1	S_2	S_3	S_4
0	1.000	0.000	0.000	0.000
1	0.600	0.400	0.000	0.000
2	0.406	0.428	0.166	0.000
3	0.289	0.385	0.296	0.030
4	0.206	0.325	0.393	0.076
5	0.152	0.290	0.413	0.145
6	0.115	0.262	0.403	0.220
7	0.098	0.217	0.399	0.286
8	0.088	0.170	0.399	0.343
9	0.063	0.175	0.375	0.387
10	0.059	0.143	0.366	0.432
11	0.048	0.118	0.366	0.468
12	0.040	0.113	0.355	0.492
13	0.042	0.104	0.336	0.518
14	0.033	0.105	0.333	0.529
15	0.031	0.104	0.315	0.550
16	0.032	0.103	0.293	0.572
17	0.025	0.091	0.311	0.573
18	0.025	0.094	0.304	0.577
19	0.029	0.083	0.298	0.590
20	0.027	0.087	0.295	0.591

4.4. Sample proportions as the estimates of true proportions

After sampling, the micro data showing the random walks are aggregated to obtain proportions for each state for each time period. These proportions, referred to as observed proportions $y_j(t)$, $j = 1, 2, ..., r$, are used to estimate true proportions $q_j(t)$, $j = 1, 2, ..., r$. That is, given $y_i(t - 1)$, $i = 1, 2, ..., r$, we assume

$$q_j(t) = \sum_i y_i(t - 1)p_{ij}, \quad \text{for } i = 1, 2, ..., r; \ t = 1, 2, ..., T. \quad (4.4.1)$$

In this section, we will show that the observed proportions are the unbiased estimates of the true proportions and therefore the error terms

$$u_j(t) = y_j(t) - q_j(t) \qquad (4.4.2)$$

have the property that $Eu_j(t) = 0, j = 1, 2, ..., r$ and $t = 1, 2, ..., T$.

Assume we have a random sample from a population whose members all exhibit an attribute 'state i' or its negative, 'not state i'. Our sample is $N(t)$ in number, and a proportion $y_i(t)$, or a number $N(t) y_i(t)$, exhibits the attribute; and consequently a proportion $1 - y_i(t)$ or a number $N(t) (1 - y_i(t))$ does not. We will assume that the population is large or that sampling is with replacement so that the probability of obtaining a 'state i' at any drawing is not affected by other drawings and therefore is a constant $q_i(t)$ at time t. The probability of obtaining $N(t) y_i(t)$ individuals in state i at time t and $N(t) (1 - y_i(t))$ individuals not in state i at time t is[2]

$$\begin{bmatrix} N(t) \\ N(t) y_i(t) \end{bmatrix} q_i(t)^{N(t) y_i(t)} (1 - q_i(t))^{N(t)(1 - y_i(t))}. \qquad (4.4.3)$$

The likelihood is then

$$L = K \prod_t \prod_i q_i(t)^{N(t) y_i(t)} (1 - q_i(t))^{N(t)(1 - y_i(t))}, \qquad (4.4.4)$$

where K denotes the constant. To obtain the ML estimator for $q_i(t)$, we can maximize $\log L$ with respect to $q_i(t)$. For the maximization of (4.4.4), we have the necessary condition

$$\frac{N(t) y_i(t)}{q_i(t)} - \frac{N(t) (1 - y_i(t))}{1 - q_i(t)} = 0, \qquad (4.4.5)$$

which gives

$$\hat{q}_i(t) = y_i(t). \qquad (4.4.6)$$

Thus, the observed proportions for each time period are the maximum likelihood estimates of the true proportions for that particular period. It is straightforward to show the ML estimators are unbiased. This can be seen by the following well known development. First, we have the relation

$$E(y_i(t)) = (1/N(t)) E(N(t) y_i(t)) = (1/N(t)) E(n_i(t)), \quad (4.4.7)$$

---

[2] If there are more than 2 states, the multinomial may be reduced to the binomial (4.4.4) when the individual is considered either to be or not to be in state i.

where $n_i(t)$ is defined to be $N(t) y_i(t)$. By definition

$$E(N(t) y_i(t)) = \sum_{n_i(t)} n_i(t) \binom{N(t)}{n_i(t)} q_i(t)^{n_i(t)} (1 - q_i(t))^{N(t) - n_i(t)}, \quad (4.4.8)$$

and the right-hand side of the equation sums to be $N(t) q_i(t)$. Using this result with (4.4.7), we have

$$E(y_i(t)) = q_i(t). \quad (4.4.9)$$

Hence, the observed proportion is an unbiased estimator. Also, it is well known to be a consistent estimator; that is,

$$\text{plim } y_i(t) = \text{plim } n_i(t)/N(t) = q_i(t). \quad (4.4.10)$$

Also, as is well known, the sample proportions have variances

$$q_i(t) (1 - q_i(t)/N(t)) \quad \text{for } i = 1, 2, ..., r; \quad t = 1, 2, ..., T, \quad (4.4.11)$$

and covariances

$$-q_i(t) q_j(t)/N(t) \quad \text{for } i \neq j \text{ and } t = 1, 2, ..., T. \quad (4.4.12)$$

These variances and covariances can be estimated consistently by

$$y_i(t) (1 - y_i(t))/N(t) \quad \text{for } i = 1, 2, ..., r; \quad t = 1, 2, ..., T \quad (4.4.13)$$

and

$$-y_i(t) y_j(t)/N(t) \quad \text{for } i \neq j \text{ and } t = 1, 2, ..., T. \quad (4.4.14)$$

These estimated variances and covariances will be employed in the generalized least squares, minimum chi-square, and maximum likelihood estimators to be developed in the chapters to follow.

4.5. Basis for gauging estimator performance

As one basis for evaluating the performance of the estimators, the root mean square error (RMSE) measure, $(\sum_{k=1}^{N} (\hat{p}_{ijk} - p_{ij})^2/N)^{\frac{1}{2}}$ where $\hat{p}_{ijk}$ is the kth sample estimate, p_{ij} is the true value and N is the number of sample estimates, will be employed. Use of this criterion means that the loss associated with an estimate, which differs from the value of the parameter being investigated, is proportional to the square root of the average of the square deviations. On the basis of prior information, there is little choice between root mean square error and mean absolute error as a dispersion measure. One point in favor of the use of the MSE measure is that it is a simple function of the bias and variance of the frequency function.

Alternatively, the performance of different estimators was evaluated by computing the size of the absolute error of the estimates from the true transition probabilities. Using this criterion, non-parametric tests involving pairwise comparisons (Siegel 1956, pp. 68–75) and ranking statistics (Siegel 1956, pp. 75–83) were used in summarizing the results. Kendall's (1961, pp. 229–239) coefficient of concordance W was used to measure the strength of the rankings.

The significance of the differences in the dispersion of the estimates about the true parameters was checked by a non-parametric test of the equality of the mean absolute errors (MAE). The hypothesis is that

$$\Pr\left(|\hat{p}_{ij} - p_{ij}| < |\tilde{p}_{ij} - p_{ij}|\right) = \tfrac{1}{2}, \qquad (4.5.1)$$

where $\hat{p}_{ij}$ and $\tilde{p}_{ij}$ denote alternative estimators and p_{ij} denotes the true parameter. The binomial test was used with this test statistic.

In addition, the Wilcoxon matched-pairs signed-ranks test was also used. In this test, the ranks of the differences are used as weights and the procedure is as follows:

$$d_k = |\hat{p}_{ijk} - p_{ij}| - |\tilde{p}_{ijk} - p_{ij}|, \qquad k = 1, 2, ..., 50, \quad (4.5.2)$$

for 50 pairs of estimates are computed and ranked without respect to sign. If d_k is zero, it is not ranked. The non-zero d_k are ranked, 1 for the smallest and 2 for the second smallest and so on. If two or more d_k have the same value, the average ranks are assigned to break the tie. After affixing to each rank the sign of the d_k which it represents, the sum of the ranks, T, with the less frequent sign is determined. The resulting statistic T is approximately normally distributed with mean $\mu_T = (N(N+1))/4$, and standard deviation $\sigma_T = \sqrt{[(N(N+1)(2N+1))/24]}$. Thus, the standardized counterpart $z = (T - \mu_T)/\sigma_T$ may be computed and compared with the tabled value of the standard normal distribution to determine the decision of acceptance or rejection of the hypothesis that the means of the two samples of estimates are the same.

4.6. Experimental results for the maximum likelihood estimator using micro data

Since complete time ordered data for 1000 individuals are available, maximum likelihood estimates of the transition probabilities were computed for each set of data in the sampling experiment in order to have a basis of comparison. As mentioned earlier, and as shown by Kendall and Stuart (1958, pp. 39–44; 93), the ML estimator for the transition probabilities based on the micro data is consistent but not generally unbiased. However, the bias tends to zero as the sample size increases and the estimator is asymptotically normally distributed.

If the data generated from the probability time path of the 1000 individuals are considered as a sample from an infinite population, then maximum likelihood estimates of the transition probabilities p_{ij} for the steps when $t = 1, 2, ..., 16$ are:

$$\dot{P} = [\dot{p}_{ij}] = \left[n_{ij} / \sum_j n_{ij} \right] = \begin{bmatrix} 0.6044 & 0.3956 & 0.0000 & 0.0000 \\ 0.0988 & 0.4930 & 0.4082 & 0.0000 \\ 0.0000 & 0.1013 & 0.6948 & 0.2039 \\ 0.0000 & 0.0000 & 0.0961 & 0.9039 \end{bmatrix}. \quad (4.6.1)$$

As might be expected, the maximum likelihood estimates based on the simulated sample yield parameters in close agreement with the true population values for the probability system (4.1.1). Application of the chi-square test[3] of the hypothesis of a stationary chain yields a value of the test statistic of 57.92 which is well below the critical value 191.75

---

[3] The chi-square test used was

$$\chi^2_{r(r-1)(t-1)} = \sum_t \sum_i \sum_j n_i (t-1) (\dot{p}_{ij}(t) - \dot{p}_{ij})^2 / \dot{p}_{ij},$$

where $\dot{p}_{ij}(t)$ is the maximum likelihood transition probability estimator for time t, $n_i (t-1)$ is the number of individuals in state i in time $t-1$ and $\dot{p}_{ij}$ are the elements of (4.6.1).

associated with the 10 percent significance level for 168 degrees of freedom.

The chi-square test[4] of the hypothesis that the estimated transition matrix (4.6.1) is consistent with (4.1.1) yields a value of 2.52, which is smaller than the tabled value of 18.55, with 12 degrees of freedom at the 10 percent significance level, and thus supports the hypothesis.

With these 1000 individuals, samples of data were drawn and maximum likelihood estimates were obtained. The experimental results for the maximum likelihood estimates for 50 samples each of size 25, 50 and 100 for $t = 1, 2, ..., 16$ are given in table 4.2.

TABLE 4.2

The means and root mean square errors for the maximum likelihood estimates

Sample size	Means				Root mean square error			
25	0.5829	0.4171	0.0000	0.0000	0.0529	0.0529	0.0000	0.0000
	0.0949	0.4998	0.4053	0.0000	0.0283	0.0656	0.0458	0.0000
	0.0000	0.1038	0.6950	0.2012	0.0000	0.0245	0.0332	0.0361
	0.0000	0.0000	0.1055	0.8945	0.0000	0.0000	0.0316	0.0316
50	0.5934	0.4066	0.0000	0.0000	0.0361	0.0361	0.0000	0.0000
	0.0999	0.4934	0.4067	0.0000	0.0265	0.0566	0.0374	0.0000
	0.0000	0.1034	0.6966	0.2000	0.0000	0.0173	0.0300	0.0265
	0.0000	0.0000	0.1060	0.8940	0.0000	0.0000	0.0245	0.0245
100	0.6077	0.3993	0.0000	0.0000	0.0224	0.0224	0.0000	0.0000
	0.0981	0.4926	0.4093	0.0000	0.0141	0.0490	0.0224	0.0000
	0.0000	0.1041	0.6975	0.1984	0.0000	0.0141	0.0200	0.0200
	0.0000	0.0000	0.1019	0.8981	0.0000	0.0000	0.0173	0.0173

As shown in this table, the means for the three different sample sizes are all close to the true population values (4.6.1) as well as to the elements of the generating matrix (4.1.1). The aggregate discrepancies of

---

[4] The chi-square test used was

$$\chi^2_{r(r-1)} = \sum_i \sum_j \left(\sum_t n_i (t-1) \right) (\dot{p}_{ij} - p_{ij})^2 / p_{ij},$$

where $\dot{p}_{ij}$ and p_{ij} are the elements of (4.6.1) and (4.1.1) respectively.

the estimates relative to (4.1.1), when expressed in terms of mean absolute errors (the sum of the absolute deviations of means of estimates from true values over all elements of the matrix), are 0.0658, 0.0486 and 0.0320, for sizes 25, 50 and 100 respectively. For samples of size 25, 50 and 100, the squared errors (the sum of squared deviations of the estimates from their respective true values, over all elements of the matrix) are about 0.0007, 0.0003 and 0.0002 respectively, and the sum of the root mean square errors (of 50 sets of estimates) over elements of the matrix are 0.4025, 0.3155 and 0.2190 for the three sample sizes. In general, as is to be expected, the estimates tend to approach the true parameter values as the sample size increases.

The frequency distributions for the non-zero transition probability estimates were compared, by means of a chi-square goodness-of-fit test, to normal distributions having means and variances equal to those of the distributions of the sample estimates. Five equal distribution intervals are determined according to their means and standard deviations so that each interval has at least five cases.[5] The chi-square values so calculated are listed in table 4.3.

TABLE 4.3

The chi-square values for the test of goodness-of-fit to the normal distributions of the maximum likelihood estimates of the transition probabilities

Elements of the matrix p_{ij}	Sample sizes		
	25	50	100
p_{11}	1.4	3.8	2.6
p_{12}	1.4	3.8	2.6
p_{21}	0.4	0.8	1.6
p_{22}	4.6	1.4	1.2
p_{23}	1.4	2.2	2.6
p_{32}	3.4	2.4	1.0
p_{33}	1.8	1.6	2.6
p_{34}	1.6	1.4	7.4
p_{43}	3.8	0.4	0.8
p_{44}	3.0	0.6	0.6

[5] For a chi-square test of goodness-of-fit to be valid, a rough rule is that no frequency count should be less than 5 (Kendall and Stuart 1961, p.440).

None of the results of the tests indicated a statistically significant departure from normality even when a level of significance as great as 10 percent was employed.[6]

4.7. Results from generated probability vector time series data

As an initial analysis, assume the probability system (4.1.1) is started in each of the four Markov states and the outcome probability vectors for $t = 1, 2, ..., 20$ for each starting state are determined (fig. 4.1). If the data contained in fig. 4.1 are considered to be four samples of time series proportion data and if the sample time period $t = 2, 3, ..., 16$ is chosen, then the resulting unrestricted least squares estimates (3.2.10) are the same as the true values of the transition probability matrix (4.1.1). However, given the regular transition matrix (4.1.1), when starting in state 1 the elements of the probability vector $y_i(t)$ exhibit the greatest variability in the initial time periods of the process and then approach the equilibrium vector as a geometrically converging sequence. These data characteristics raise a question relative to the impact of the choice of the length and location of the sample period on the unrestricted least squares estimates of (4.1.1). To get some idea of the properties of unrestricted least squares estimates in this situation, estimates were obtained by using observations on the probability vector for $t = 2, 3, ..., 6$, etc. The results are given in table 4.4.

In table 4.4, for the sample period $t = 2, 3, ..., 6$, the least squares estimator (3.2.10) yielded estimates very close to the true parameters. However, when the sample period $t = 8, 9, ..., 12$ was used, the unrestricted least squares estimates missed the mark badly, and the condition of non-negative probabilities was violated. This result was brought about by the high degree of collinearity between the X variables and the near singularity of the $X'X$ matrix. These characteristics may be encountered frequently in actual economic time series data.

---

[6] The critical value for chi-square with four degrees of freedom, at a significance level of 10 percent, is 7.78.

TABLE 4.4

The classical least squares estimates of the transition matrix from different portions of the aggregate data for a Markov process

Time period	Unrestricted least squares	Restricted least squares
$t = 2, 3, \ldots, 6$	$\begin{bmatrix} 0.6000 & 0.3996 & -0.0006 & 0.0010 \\ 0.1000 & 0.5007 & 0.4009 & -0.0016 \\ 0.0000 & 0.0993 & 0.6991 & 0.2016 \\ -0.0003 & 0.0013 & 0.1015 & 0.8995 \end{bmatrix}$	$\begin{bmatrix} 0.6001 & 0.3999 & 0.0000 & 0.0000 \\ 0.0999 & 0.5001 & 0.4000 & 0.0000 \\ 0.0000 & 0.0999 & 0.6999 & 0.2002 \\ 0.0000 & 0.0004 & 0.1003 & 0.8993 \end{bmatrix}$
$t = 3, 4, \ldots, 7$	$\begin{bmatrix} 0.5926 & 0.4003 & 0.0020 & 0.0051 \\ 0.1084 & 0.4997 & 0.3978 & -0.0059 \\ -0.0047 & 0.1000 & 0.7011 & 0.2036 \\ 0.0038 & 0.0004 & 0.0993 & 0.8965 \end{bmatrix}$	$\begin{bmatrix} 0.6006 & 0.3998 & 0.0005 & 0.0000 \\ 0.0994 & 0.5013 & 0.3993 & 0.0000 \\ 0.0000 & 0.0094 & 0.7004 & 0.2002 \\ 0.0005 & 0.0007 & 0.0997 & 0.8992 \end{bmatrix}$
$t = 5, 6, \ldots, 9$	$\begin{bmatrix} 0.5858 & 0.3751 & -0.0562 & 0.1060 \\ 0.1119 & 0.5203 & 0.4488 & -0.0906 \\ -0.0333 & 0.0953 & 0.6859 & 0.2250 \\ 0.0007 & 0.0001 & 0.1031 & 0.8954 \end{bmatrix}$	$\begin{bmatrix} 0.5990 & 0.3775 & 0.0199 & 0.0036 \\ 0.1005 & 0.5182 & 0.3813 & 0.0000 \\ 0.0000 & 0.0956 & 0.7065 & 0.1977 \\ 0.0000 & 0.0000 & 0.0979 & 0.9021 \end{bmatrix}$
$t = 8, 9, \ldots, 12$	$\begin{bmatrix} 0.3179 & 0.6159 & -0.4783 & 0.2704 \\ 0.3226 & 0.3260 & 0.7974 & -0.2206 \\ -0.0539 & 0.1434 & 0.5953 & 0.2567 \\ 0.0078 & -0.0065 & 0.1178 & 0.8907 \end{bmatrix}$	$\begin{bmatrix} 0.5945 & 0.4055 & 0.0000 & 0.0000 \\ 0.1023 & 0.4963 & 0.4014 & 0.0000 \\ 0.0000 & 0.1004 & 0.6993 & 0.2002 \\ 0.0000 & 0.0002 & 0.1000 & 0.8998 \end{bmatrix}$

4.8. Results from the sampling experiment macro data

a. Unrestricted least squares

Making use of experimental data generated from a population of 1000 individuals, information on the proportion of individuals in each state for time period 1 through 16 was used to obtain the unrestricted least squares estimates of the transition probabilities. For each case, the non-negativity condition on the set of transition probabilities was violated. The mean and root mean square error statistics for the unrestricted estimates for each level of 50 problems are presented in table 4.5.

TABLE 4.5

Means and root mean square errors for the unrestricted least squares estimates

Sample size	Means				Root mean square error			
25	0.5165	0.5089	0.0180	−0.0434	0.1583	0.2228	0.2134	0.1407
	0.1504	0.3866	0.4466	0.0164	0.1437	0.2170	0.2541	0.1608
	−0.0104	0.1408	0.6067	0.2629	0.1127	0.1272	0.1795	0.1265
	−0.0001	0.0007	0.1586	0.8408	0.0802	0.0973	0.1503	0.1099
50	0.5666	0.4547	−0.0344	0.0131	0.1041	0.1656	0.1793	0.1112
	0.1471	0.4067	0.4972	−0.0410	0.1179	0.1758	0.2095	0.1480
	−0.0412	0.1701	0.6127	0.2584	0.0974	0.1457	0.1434	0.1034
	0.0257	−0.0310	0.1506	0.8547	0.0621	0.1007	0.1054	0.0842
100	0.5480	0.4857	−0.0218	−0.0119	0.1115	0.1526	0.1845	0.1268
	0.1724	0.2553	0.4719	0.0004	0.1494	0.2017	0.2295	0.1615
	−0.0350	0.1806	0.6501	0.2043	0.0824	0.1235	0.1137	0.0858
	0.0144	−0.0260	0.1114	0.9002	0.0411	0.0688	0.0655	0.0513

For each sample size, three or four of the means violate the non-negativity condition. In addition, they show up very badly when their means and root mean square errors are compared with the maximum likelihood results from *micro* data as shown in the previous section. As expected, the root mean square errors decrease as the sample size increases. When the root mean square errors are summed over all elements of the transition matrix without weighting the individual elements, the aggregated errors are 2.49, 2.05 and 1.95 for the estimates based on samples of sizes 25, 50 and 100 respectively.

In order to test empirically the distributional properties of the estimates, ninety samples of size 100 were drawn and the frequency distributions of each of the sixteen estimates were compared by means of a chi-square test to normal distributions having means and variances equal to those computed from the estimates. Of the sixteen estimates, only one (p_{13}) showed a significant deviation from normality at the 5 percent significance level.

To test for significant differences in the central tendency of each of the estimates from the population parameter values, use was made of the t test with the statistic (Johnston [1963, p. 118]):

$$\frac{\tilde{p}_{ij} - p_{ij}}{\sqrt{[(\text{SSE}/(Tr - r^2))\, a_{ii}]},} \tag{4.8.1}$$

where SSE denotes sum of square errors and a_{ii} denotes the ith diagonal element of the matrix $(X'X)^{-1}$. For the unrestricted least squares estimates, these tests resulted in no significant deviations from the true values.

b. Restricted least squares

The means and root mean square error statistics for the restricted least squares estimator (see § 3.3) for each set of 50 estimates are presented in table 4.6.

The restricted least squares estimates of the transition probabilities appear much superior to the unrestricted estimates given in table 4.5. Superiority is indicated in that the means, of course, are non-negative and the root mean square errors are absolutely smaller than those of the corresponding unrestricted estimates.

To examine the sample distribution of the restricted least squares estimates, the Smirnov–Kolomogorov D statistic (Lindgren 1962, p. 300) is computed for the test of goodness-of-fit to the normal distribution. The D statistic is defined to be

$$Dn = \sup_{\text{all } x} |Fn(x) - F(x)|, \tag{4.8.2}$$

TABLE 4.6

Means and root mean square errors for the restricted least squares estimates

Sample size	Means				Root mean square error			
25	0.4992	0.4272	0.0723	0.0013	0.1521	0.1513	0.1369	0.0058
	0.1315	0.4593	0.3943	0.0149	0.1011	0.1603	0.1406	0.0291
	0.0160	0.0992	0.6344	0.2504	0.0307	0.0649	0.1324	0.0921
	0.0051	0.0262	0.1361	0.8326	0.0143	0.0407	0.1123	0.1149
50	0.5680	0.3933	0.0386	0.0001	0.0881	0.1159	0.0842	0.0000
	0.1080	0.4806	0.4020	0.0094	0.0696	0.1155	0.0888	0.0192
	0.0069	0.1031	0.6562	0.2338	0.0168	0.0560	0.0965	0.0684
	0.0039	0.0174	0.1247	0.8540	0.0086	0.0364	0.0744	0.0807
100	0.5657	0.3953	0.0390	0.0000	0.0868	0.1147	0.0840	0.0000
	0.1224	0.4754	0.3959	0.0063	0.0712	0.1199	0.0912	0.0145
	0.0043	0.1060	0.6875	0.2022	0.0099	0.0393	0.0629	0.0404
	0.0017	0.0113	0.0929	0.8941	0.0049	0.0232	0.0459	0.0432

where $Fn(x)$ is the sample distribution which has the form

$$Fn(x) = j/N, \quad \text{for } X(j) < x < X(j+1), \quad j = 0, 1, ..., N, \quad (4.8.3)$$

and $F(x)$ is the assumed normal distribution function.

Fifty estimates each of sample size 50 are used for the test. The D statistics for the estimates of the non-zero true parameters are

$$\begin{bmatrix} 0.0857 & 0.0671 & ----- & ----- \\ 0.1102 & 0.0093 & 0.0880 & ----- \\ ----- & 0.0663 & 0.1336 & 0.0759 \\ ----- & ----- & 0.0842 & 0.1108 \end{bmatrix}, \quad (4.8.4)$$

in which none of them is larger than the tabled value 0.17 for sample size 50 at the 10 percent significance level. Thus, we cannot reject the hypothesis that the estimates are normally distributed. As for the estimates of zero parameters, they are truncated at zero. Their characteristics are listed in table 4.7.

In table 4.7, at least 60 percent of the estimates exactly hit the target values. Extreme values are not large except for the estimates of p_{13} and p_{42}. The extreme estimate for p_{13} is unusually large but the next largest estimate is only 0.1909. The estimates for p_{14} are almost perfect and only one estimate with a value of 0.0049 misses the target.

TABLE 4.7

Some characteristics of the estimates of the zero true parameters by the restricted least squares

Variable	No. of estimates at zero	No. of positive estimates	Extreme value	Mean	Standard deviation
p_{13}	30	20	0.4109	0.0386	0.0748
p_{14}	49	1	0.0049	0.0001	0.0007
p_{24}	35	15	0.0583	0.0084	0.0167
p_{31}	38	12	0.0611	0.0069	0.0153
p_{41}	35	15	0.0302	0.0039	0.0076
p_{42}	33	17	0.1448	0.0174	0.0320

4.9. An application

In addition to the artificial data generated for the sampling experiments, let us also consider a sample of real data, pertaining to brand changes for cigarette smokers, as one illustration of the applicability of transition probability estimators.

a. The brand change problem

In a recent study, Telser (1963) investigated the time ordered brand behavior pattern for cigarette smokers. Time ordered brand data relating to each individual smoker were not available. Only annual data, giving the sales in billions of cigarettes for the three leading brands were available. These were used to provide the figures shown in table 4.8.

Given this information, Telser assumed that the market shares are obtained from a fixed sample of smokers and that the behavior pattern

TABLE 4.8
Market shares of Camels, Lucky Strike and Chesterfield,
and total sales of cigarettes

Year	Camels	Lucky Strike	Chester-field	Total sales (billions)
1925	0.5056	0.2028	0.2916	67.9
26	0.4879	0.1899	0.3222	77.9
27	0.4504	0.2236	0.3260	85.3
28	0.4068	0.3039	0.2893	90.2
29	0.3637	0.3616	0.2747	102.0
1930	0.3365	0.4118	0.2517	104.0
31	0.3311	0.4425	0.2264	100.4
32	0.2936	0.4498	0.2566	81.5
33	0.2794	0.4008	0.3198	91.6
34	0.3418	0.3301	0.3281	97.6
1935	0.3867	0.3013	0.3120	101.9
36	0.4074	0.2906	0.3020	113.9
37	0.4084	0.2949	0.2967	116.9
38	0.3842	0.3195	0.2963	113.8
39	0.3746	0.3358	0.2896	114.2
1940	0.3708	0.3500	0.2792	120.0
41	0.3579	0.2653	0.2768	135.5
42	0.3527	0.3851	0.2622	154.5
43	0.3276	0.3875	0.2849	175.5

Source: William H. Nicholls, *Price Policies in the Cigarette Industry*, Nashville, Vanderbilt University Press, 1951, pp. 64, 94, 142.

of the smokers could be reflected by a stationary first order Markov process.

b. *Results for the restricted and unrestricted estimators*

Given the assumption that the data were generated from a first order Markov process, Telser made use of the unrestricted least squares transition probability estimator (3.2.10) to obtain the following estimated transition probability matrix:

$$\tilde{P} = [\tilde{p}_{ij}] = \begin{bmatrix} 0.5454 & 0.3574 & 0.0972 \\ -0.0744 & 0.9654 & 0.1090 \\ 0.6489 & -0.3949 & 0.7460 \end{bmatrix}. \qquad (4.9.1)$$

Two of the elements are negative and not acceptable as transition probabilities. If non-negative restrictions are imposed in the estimation procedure, then the restricted least squares estimates obtained via the quadratic programming procedure of § 3.3 are:

$$\tilde{P}^c = [\tilde{p}_{i,j}^c] = \begin{bmatrix} 0.6686 & 0.1423 & 0.1891 \\ 0.0000 & 0.8683 & 0.1317 \\ 0.4019 & 0.0000 & 0.5981 \end{bmatrix}. \tag{4.9.2}$$

Smoker behavior defined by the probability system (4.9.2) reflects a high degree of brand loyalty from one year to the next and for the brands represented by states 2 and 3, a willingness to switch to only one of the other brands.

CHAPTER 5

WEIGHTED INEQUALITY RESTRICTED LEAST SQUARES ESTIMATORS

In developing a restricted least squares estimator in ch. 3, we ignored the correct specification for the covariance disturbance matrix Σ (3.2.8) and proceeded as if Σ took the form of a scalar times a $(Tr \times Tr)$ identity matrix. Madansky (1959) has noted that this assumption is invalid. With a model involving proportions, heteroscedasticity is present and thus the unweighted restricted least squares estimator, which does not take account of heteroscedasticity, is inefficient. Given this situation, one might consider applying Aitken's generalized least squares approach (1934) to obtain the following minimum variance linear unbiased estimator:

$$\tilde{\tilde{p}} = (X'\Sigma^{-1}X)^{-1}X'\Sigma^{-1}y. \tag{5.0.1}$$

It should be noted, however, that for the model considered in this study, Σ is a singular matrix and thus Σ^{-1} does not exist. Given the singularity of Σ, one way to deal with the problem of heteroscedasticity is to introduce a $(Tr \times Tr)$ non-singular 'weight' matrix to replace Σ in (5.0.1).[1]

In this chapter, several weight matrices will be employed to construct estimators and the properties of these estimators will be compared with those of the unweighted least squares estimator by means of sampling experiments. This will provide some information on the extent to which the estimators' properties are sensitive to alterations in weighting patterns.

---

[1] The generalized inverse method has been suggested by Chipman (1964), Rao (1962) and others as one way to solve this type of singularity problem. However, our problem not only has a singular disturbance covariance matrix but also has linear dependence among the equations to be estimated. The generalized inverse method as it applies to this problem is discussed in appendix A.

5.1. The statistical model

As specified in ch. 3, the statistical model is

$$y = Xp + u, \tag{5.1.1}$$

with

$$Eu = 0 \tag{5.1.2}$$

and

$$Euu' = \Sigma, \tag{5.1.3}$$

where y and u are $(rT \times 1)$ column vectors, X is an $(rT \times r^2)$ block diagonal matrix, p is an $(r^2 \times 1)$ column vector and Σ is an $(rT \times rT)$ non-diagonal, singular matrix.

Reformulating the model within the generalized least squares framework (Aitken 1934), we take into account the weight of each observation (or each time period) under the assumption of heteroscedasticity. Applying an $(rT \times rT)$ weight matrix H to (5.1.1), we have

$$Hy = HXp + Hu, \tag{5.1.4}$$

and the weighted least squares estimate of p is then

$$\tilde{p} = (X'H'HX)^{-1} X'H'Hy. \tag{5.1.5}$$

Given this result, it is to the problem of finding an 'appropriate' H matrix that we now turn.

5.2. Weighted restricted least squares

Since Σ (5.1.3) is singular, there exists no matrix H that will transform the heteroscedastic disturbances into homoscedastic disturbances. However, there are weight matrices for the observations which will define a weighted least squares estimator which may be more efficient than the classical unweighted least squares estimator in (3.2.10). In starting the search, we shall suggest some weight matrices which will give the same weight within each equation but different weights among equations. Then we will try different weights for different observations within and among equations.

For the case when weights are different among equations but the same weights are given within the equation, the unrestricted weighted least squares estimates will be the same as the unrestricted unweighted least squares estimates. That is, if $H'H$ has the form

$$H'H = \hat{\Sigma} \otimes I, \qquad (5.2.1)$$

where $\hat{\Sigma}$ is an $(r \times r)$ matrix, $\otimes$ denotes the Kronecker product and I a $(T \times T)$ identity matrix, then

$$(X'H'HX)^{-1} X'H'Hy = (X'X)^{-1} X'y, \qquad (5.2.2)$$

since X is block diagonal and all blocks are identical.

The result in (5.2.2) seems to discourage the use of weights. However, when constraints are imposed on the parameters, the restricted weighted least squares estimator may be different from the unrestricted estimator because the local minima of the squared error loss function may be affected by the weights.

For the more general case, if different weights are given to different equations and also to different observations within equations, the unrestricted weighted least squares estimator may violate the non-negativity and row sum constraints on the transition probabilities. Let the matrix H be an $(rT \times rT)$ block diagonal matrix as follows:

$$H = \begin{bmatrix} a_1 d_1 & & & \\ & a_2 d_2 & & \\ & & \ddots & \\ & & & a_r d_r \end{bmatrix}, \qquad (5.2.3)$$

where a_i, $i = 1, 2, ..., r$, are scalars and d_i, $i = 1, 2, ..., r$, are $(T \times T)$ diagonal matrices. The weight matrix is then

$$H'H = \begin{bmatrix} a_1^2 d_1' d_1 & & & \\ & a_2^2 d_2' d_2 & & \\ & & \ddots & \\ & & & a_r^2 d_r' d_r \end{bmatrix}. \qquad (5.2.4)$$

Although in this case, the p_j may be estimated independently of the p_i $(i \neq j)$, the unrestricted least square estimates may violate the unit row sum constraints condition. Using the (2×2) case as an example, the row sums of the estimates of transition probabilities are

$$\tilde{p}_1 + \tilde{p}_2 = (X'd_1'd_1 X)^{-1} X'd_1'd_1 y_1 + (X'd_2'd_2 X)^{-1} X'd_2'd_2 y_2, \qquad (5.2.5)$$

which may not be equal to η_2 unless d_i has a special form. Thus, only the restricted estimator (i.e., an estimator which satisfies both (1.1.5) and (1.1.6)) is meaningful.

5.3. Some alternative weights

In order to mitigate the impact of heteroscedasticity between equations and/or within equations, some alternative weights that may be employed to derive weighted restricted least squares estimates are:

(1) Different weights among equations and the same weights within equations, i.e., $d_i = I$ and one of the following a_i's:

(a) The inverse of the estimate of the ith equation disturbance variance σ^2, i.e.,

$$a_i = (T - r)/[(y_i - X\tilde{p}_i)' (y_i - X\tilde{p}_i)], \qquad (5.3.1)$$

where $\tilde{p}_i$ is the classical unrestricted estimate.

(b) The inverse of the average of the proportion for each y_i variable, i.e.,

$$a_i = T/ \left(\sum_t y_i(t) \right). \qquad (5.3.2)$$

(c) The inverse of the product of the average proportion in state i and the average proportion not in state i, i.e.,

$$a_i = 1 \left| \left[\frac{\sum_t y_i(t)}{T} \frac{1 - \sum_t y_i(t)}{T} \right] \right. . \qquad (5.3.3)$$

(2) Different weights among and within equations.

The disturbance variance, as an approximation, is assumed to be proportional to the true proportion in the following way:

$$\sigma_j^2(t) \propto \begin{cases} q_j(t)/N(t) & \text{if } q_j(t) \leq 0.5 \\ (1 - q_j(t))/N(t) & \text{if } q_j(t) \geq 0.5 \end{cases}. \qquad (5.3.4)$$

This means that the disturbance variance is a maximum at $q_j(t) = \frac{1}{2}$ and is zero at the two boundary points zero and unity. Since most of the observed proportions are less than 0.5 in our experimental model and also for simplicity, we simply construct the weight matrix as

$$a_j^2 d_j' d_j = \begin{bmatrix} N(1)/q_j(1) & & & \\ & N(2)/q_j(2) & & \\ & & \ddots & \\ & & & N(T)/q_j(T) \end{bmatrix}, \quad (5.3.5)$$

for all j from 1 to r, where $q_j(t)$ may be substituted for by the observed proportions $y_j(t)$.

Developed in this way, the choice of these weights is somewhat arbitrary. Other weights may be employed and examples of these are contained in the works of Madansky (1959) and Theil and Rey (1966). Our objective now is to use data from the sampling experiment to find if these weights will result in estimates which have a smaller variance than the classical restricted least squares estimates.

In later chapters, we will consider various formulations of the restricted estimator weighted by (5.2.4) in the form of (5.3.5). The sampling results using the weights (5.3.1), (5.3.2) and (5.3.3) are given in the following section.

5.4. Results from sampling experiment

Proceeding as in ch. 4, fifty estimates for samples of size 25, 50 and 100 are obtained by the weighted restricted least squares estimates procedure. The estimate means and root mean square errors are also computed for various sample sizes.

a. Weighted by average proportion of i-th state

When use is made of (5.3.2), the reciprocal of the average proportion for the ith state in the sample period, the means and root mean square errors for 50 estimates of size 25, 50 and 100 given in table 5.1 result.

The means of the estimates appear to be closer to the true population parameters and the root mean square errors smaller than for the unweighted restricted least squares. The sum of the mean absolute deviations of the estimates from the population parameters for estimates of size 25, 50 and 100 are 0.5025, 0.2660 and 0.1611, respectively. These results show that the new estimates have smaller MSE's and MAD's than

TABLE 5.1
Weighted restricted least squares estimates with the average proportion in the ith state as weights

Sample size	Means				Root mean square error			
25	0.5193	0.4181	0.0613	0.0013	0.1372	0.1388	0.1176	0.0057
	0.1213	0.4632	0.3998	0.0157	0.0954	0.1565	0.1294	0.0297
	0.0144	0.1003	0.6371	0.2482	0.0286	0.0658	0.1291	0.0894
	0.0066	0.0267	0.1325	0.8342	0.0162	0.0452	0.1090	0.1116
50	0.5801	0.3851	0.0348	0.0000	0.0833	0.1120	0.0775	0.0000
	0.1019	0.4880	0.4008	0.0093	0.0694	0.1115	0.0838	0.0192
	0.0061	0.1021	0.6588	0.2330	0.0157	0.0566	0.0950	0.0674
	0.0052	0.0177	0.1224	0.8547	0.0106	0.0369	0.0731	0.0795
100	0.5757	0.3923	0.0320	0.0000	0.0808	0.1054	0.0709	0.0000
	0.1161	0.4773	0.4005	0.0061	0.0672	0.1122	0.0779	0.0141
	0.0036	0.1063	0.6881	0.2020	0.0088	0.0398	0.0599	0.0397
	0.0025	0.0116	0.0913	0.8946	0.0057	0.0237	0.0448	0.0422

those associated with the unweighted restricted least squares estimator. They also have the property that when sample size increases, the estimates get closer to the true parameter.

b. Weighted by an estimate of the i-th equation disturbance variance

When the inverse of the estimate of the disturbance variance for each equation (5.3.1) is used as weight to that equation, then the means and root mean square errors for 50 estimates of sample size 25, 50 and 100 given in table 5.2 result.

The estimates of table 5.2 appear superior to the unweighted estimates of ch. 4 since the root mean square errors for each sample size are smaller. The sum of the mean absolute deviations of the estimates from the population parameters for estimates of size 25, 50 and 100 are 0.4755, 0.2472 and 0.1519, respectively.

c. Weighted by product of average proportions in state i

When the estimation equations are weighted by the inverse of the product of the average sample proportion in state i and the average propor-

TABLE 5.2
Weighted restricted least squares estimates with equation disturbance variances as weights

Sample size	Means				Root mean square error			
25	0.5238	0.4129	0.0618	0.0015	0.1343	0.1356	0.1187	0.0060
	0.1189	0.4668	0.4005	0.0138	0.0946	0.1549	0.1316	0.0284
	0.0139	0.1010	0.6375	0.2476	0.0279	0.0655	0.1268	0.0880
	0.0064	0.0251	0.1304	0.8381	0.0157	0.0434	0.1032	0.1066
50	0.5828	0.3811	0.0360	0.0001	0.0819	0.1130	0.0797	0.0000
	0.1001	0.4926	0.3990	0.0083	0.0674	0.1119	0.0845	0.0177
	0.0060	0.1008	0.6624	0.2308	0.0158	0.0534	0.0910	0.0642
	0.0051	0.0164	0.1193	0.8592	0.0104	0.0352	0.0703	0.0751
100	0.5799	0.3844	0.0357	0.0000	0.0806	0.1119	0.0748	0.0000
	0.1126	0.4879	0.3945	0.0050	0.0667	0.1160	0.0833	0.0126
	0.0037	0.1026	0.6922	0.2015	0.0087	0.0377	0.0573	0.0378
	0.0028	0.0113	0.0892	0.8967	0.0063	0.0229	0.0430	0.0406

TABLE 5.3
Weighted restricted least squares estimates with the product of state proportions as weights

Sample size	Means				Root mean square error			
25	0.5161	0.4180	0.0646	0.0013	0.1392	0.1416	0.1236	0.0058
	0.1229	0.4638	0.3980	0.0153	0.0962	0.1569	0.1330	0.0293
	0.0146	0.0999	0.6362	0.2493	0.0290	0.0653	0.1301	0.0905
	0.0063	0.0264	0.1339	0.8334	0.0159	0.0446	0.1103	0.1130
50	0.5780	0.3859	0.0361	0.0000	0.0838	0.1130	0.0795	0.0000
	0.1028	0.4869	0.4010	0.0093	0.0692	0.1123	0.0852	0.0191
	0.0063	0.1022	0.6582	0.2333	0.0161	0.0563	0.0954	0.0677
	0.0049	0.0176	0.1232	0.8543	0.0102	0.0367	0.0736	0.0800
100	0.5738	0.3918	0.0344	0.0000	0.0818	0.1083	0.0755	0.0000
	0.1172	0.4779	0.3988	0.0061	0.0680	0.1144	0.0824	0.0142
	0.0039	0.1060	0.6880	0.2021	0.0091	0.0396	0.0609	0.0398
	0.0022	0.0115	0.0920	0.8943	0.0055	0.0235	0.0453	0.0425

tion not in state i (5.3.3), the restricted estimates given in table 5.3 result.

Again, the weighted restricted least squares estimates appear superior to the unweighted estimates. As compared to the other weighted estimates, the root mean square errors are slightly larger. However, when comparisons among the three kinds of weights are given in table 5.4, it is quite difficult to say which weight is absolutely better than the other.

TABLE 5.4

Aggregate root mean square errors from weighted and unweighted restricted estimators

Sample size	Unweighted	Weighted by the inverse of		
		Means	Product of state proportions	Disturbance variances
25	1.4795	1.4052	1.3812	1.4243
50	1.0191	0.9915	0.9715	0.9981
100	0.8520	0.7931	0.8002	0.8108

The weighted estimates appear superior in all cases to the unweighted estimates. Among alternative weights, the inverse of the product of state proportions appears inferior to the others. For a sample size of 25 or 50, the inverse of the disturbance variance has smaller root mean square errors than the inverse of the means. However, when the sample size is large the mean proportions provide better weights.

In regard to the distribution of the estimates, 50 estimates of size 50 weighted by the estimated disturbance variances are used for the Kolmogorov–Smirnov D test for the goodness-of-fit to the normal distribution. The resulting D statistics for the non-zero parameters are

$$\begin{bmatrix} 0.1004 & 0.0806 & ----- & ----- \\ 0.0720 & 0.0705 & 0.0951 & ----- \\ ----- & 0.0683 & 0.1301 & 0.0763 \\ ----- & ----- & 0.0663 & 0.0894 \end{bmatrix}. \qquad (5.4.1)$$

All the figures in (5.4.1) are less than the acceptance limit 0.17, for samples of size 50, at the 10 percent significant level. Thus, we may conclude that the estimates are normally distributed.

The results for the estimates of the zero value parameters are listed in Table 5.5. When Table 5.5 is compared with table 4.7 in ch. 4, it is apparent that except for p_{41}, the distributions of these estimates have smaller variances, with the means closer to zero.

TABLE 5.5

Some characteristics of the estimates of the zero value true parameters by restricted least squares weighted by the inverse of the disturbance variances

Variable	No. of estimates at zero	No. of positive estimates	Extreme value	Means	Standard deviation
p_{13}	30	20	0.3807	0.0360	0.0711
p_{14}	49	1	0.0037	0.0001	0.0005
p_{24}	35	15	0.0547	0.0083	0.0156
p_{31}	38	12	0.0590	0.0060	0.0146
p_{41}	33	17	0.0317	0.0051	0.0089
p_{42}	33	17	0.1415	0.0164	0.0311

As a whole, the weighted estimates have smaller root mean square errors and also maintain the property of consistency.

The sampling results for the restricted least squares estimator involving different weights within and among equations is given in the following chapter.

5.5. Results for the brand change problem

For the Telser smoker problem introduced in §4.9, the weighted restricted least squares estimator (weighted by the inverse of the mean proportions) is

$$\tilde{P} = [\tilde{p}_{ij}] = \begin{bmatrix} 0.6790 & 0.1416 & 0.1794 \\ 0.0000 & 0.8691 & 0.1309 \\ 0.3881 & 0.0000 & 0.6119 \end{bmatrix} . \, . \qquad (5.5.1)$$

This result agrees quite closely with that obtained for the unweighted restricted least squares estimator.

CHAPTER 6

A GENERALIZED LEAST SQUARES ESTIMATOR

The singularity of the variance-covariance matrix of the disturbances for the Markov model was noted in previous chapters. Since the matrix is singular, Aitken's generalized least squares method cannot be employed. Close examination, however, reveals that the singularity is due to redundant variables and thus the situation becomes tractable when the redundant variables are deleted. This chapter is devoted to the exposition and consequences of this proposition.

6.1. Non-spherical disturbances

Let us restate the statistical model presented in ch. 3 and write the set of r Markov chain relations as the single relation:

$$
\begin{bmatrix} y_1 \\ y_2 \\ \vdots \\ y_r \end{bmatrix} = \begin{bmatrix} X_1 & & & \\ & X_2 & & \\ & & \ddots & \\ & & & X_r \end{bmatrix} \begin{bmatrix} p_1 \\ p_2 \\ \vdots \\ p_r \end{bmatrix} + \begin{bmatrix} u_1 \\ u_2 \\ \vdots \\ u_r \end{bmatrix}
\tag{6.1.1}
$$

or more compactly as

$$
y = Xp + u,
\tag{6.1.2}
$$

with $X_1 = X_2 = \cdots = X_r$,

$$
E(u) = 0
\tag{6.1.3}
$$

and

$$
E(uu') = \Sigma,
\tag{6.1.4}
$$

where Σ is a $((Tr) \times (Tr))$ covariance matrix. We assume that u is at least contemporaneously independent of X and thus y and u have the same covariance matrix. It is further assumed that, y, a set of proportions, is generated from a multinomial distribution which has means $q_j(t)$, variances $q_j(t)[1 - q_j(t)]/N(t)$, and covariances $-q_i(t)$

$q_j(t)/N(t)$. Therefore, u has a zero mean vector and the following $(rT \times rT)$ covariance matrix:

$$
\Sigma = \begin{bmatrix}
\Sigma_{11} & \Sigma_{12} & \cdots & \Sigma_{1r} \\
\Sigma_{21} & \Sigma_{22} & \cdots & \Sigma_{2r} \\
\vdots & & \ddots & \vdots \\
\Sigma_{r1} & \Sigma_{r2} & \cdots & \Sigma_{rr}
\end{bmatrix}, \tag{6.1.5}
$$

where the sub-matrices of (6.1.5) are $(T \times T)$ with

$$
\Sigma_{ij} = \begin{bmatrix}
-\dfrac{q_i(1)\, q_j(1)}{N(1)} & & & \\
& -\dfrac{q_i(2)\, q_j(2)}{N(2)} & & \\
& & \ddots & \\
& & & -\dfrac{q_i(T)\, q_j(T)}{N(T)}
\end{bmatrix} \quad \text{for } i \neq j, \tag{6.1.6}
$$

and

$$
\Sigma_{ii} = \begin{bmatrix}
\dfrac{q_i(1)\, [1 - q_i(1)]}{N(1)} & & & \\
& \dfrac{q_i(2)\, [1 - q_i(2)]}{N(2)} & & \\
& & \ddots & \\
& & & \dfrac{q_i(T)\, [1 - q_i(T)]}{N(T)}
\end{bmatrix}. \tag{6.1.7}
$$

This covariance structure has the following properties:

(1) Each equation of the multivariate model (6.1.1) has heteroscedastic disturbances.

(2) Among equations the disturbances are interdependent within the same sample period.

(3) The disturbances are not autocorrelated.

Property (1) is the result of the multinomial distribution under the Lexis scheme.[1] Property (2) is a consequence of the interdependence of

---

[1] For a discussion of the multinomial distribution under the Lexis scheme, see Reitz (1934).

the proportion data. Property (3) is based on the fact that although the observations at each time period are from the same parent population, they may not be correlated with the previous observations since the number of observations at each time period $N(t)$ may be different. In any event, the whole system (6.1.1) has non-spherical disturbances.

6.2. Redundant parameters and the reduced model

The covariance matrix (6.1.5) is symmetric and singular. It is singular because in each row, the row-sum is zero due to the fact that

$$\sum_{i=1}^{r} q_i(t) = 1, \qquad \text{for all } t. \tag{6.2.1}$$

Thus, Aitken's generalized least squares cannot be applied directly to (6.1.1) to obtain the best linear unbiased estimator.

Examining the model (6.1.1), note that the elements of the y_i's and X_i's are proportions, the p_i's are probabilities and the u_i's are inter-related over equations by (6.1.5). Thus, the equations in (6.1.1) are linearly dependent, and if $r-1$ of the equations in (6.1.1) are known, the remaining equation is also known by relation (6.2.1). In other words, the identity (6.2.1) repeats information for the model defined by (6.1.1) and there exists a set of redundant parameters.

Since there are only $r-1$ independent observation vectors y_j and $r-1$ independent parameter vectors p_j, and since the identity (6.2.1) cannot be neglected, we may delete one of the equations from the model (6.1.1).[2] Without loss of generality, we may delete the last equation since these equations are free of the order of permutation. Thus, the reduced model can be written as

$$\begin{bmatrix} y_1 \\ y_2 \\ \vdots \\ y_{r-1} \end{bmatrix} = \begin{bmatrix} X_1 & & & \\ & X_2 & & \\ & & \ddots & \\ & & & X_{r-1} \end{bmatrix} \begin{bmatrix} p_1 \\ p_2 \\ \vdots \\ p_{r-1} \end{bmatrix} + \begin{bmatrix} u_1 \\ u_2 \\ \vdots \\ u_{r-1} \end{bmatrix} \tag{6.2.2}$$

[2] This deletion problem is parallel to the case of finding the equilibrium vector for a regular Markov chain (Kemeny et al. 1958, p. 394).

or more compactly as

$$y_* = X_* p_* + u_*, \tag{6.2.3}$$

where the asterisks denote the subset of observations, parameters and disturbances or the whole sub-system of relations. The disturbance u_* then has the following specifications:

$$Eu_* = 0 \tag{6.2.4}$$

and

$$Eu_* u'_* = \Sigma_*, \tag{6.2.5}$$

where Σ_* is a $((r-1)\,T \times (r-1)\,T)$ non-singular matrix, a submatrix of Σ.

6.3. Existence of the inverse of the disturbance covariance matrix

The disturbance covariance matrix Σ_* is a submatrix of (6.1.5) with the last row-blocks ($\hat{\Sigma}_{rj}, j = 1, 2, ..., r$) and the last column-blocks ($\hat{\Sigma}_{ir}, i = 1, 2, ..., r$) deleted. Its inverse Σ_*^{-1} may serve as a weight matrix in forming a weighted least squares estimator. To show the inversion of Σ_* in a simple way, we need some new mathematical notation.

Define $V_*(t)$ to be a $((r-1) \times (r-1))$ cross-section matrix of Σ_* at time t in the form

$$V_*(t) = \frac{1}{N(t)} \begin{bmatrix} q_1(t)\,[1 - q_1(t)] & -q_1(t)\,q_2(t) & \cdots & -q_1(t)\,q_{r-1}(t) \\ -q_2(t)\,q_1(t) & q_2(t)\,[1 - q_2(t)] & \cdots & -q_2(t)\,q_{r-1}(t) \\ \vdots & \vdots & & \vdots \\ -q_{r-1}(t)\,q_1(t) & -q_{r-1}(t)\,q_2(t) & \cdots & q_{r-1}(t)\,[1 - q_{r-1}(t)] \end{bmatrix}. \tag{6.3.1}$$

Also, we define a time-ordered Kronecker expansion of $V_*(t)$ as

$$V_*(t) \otimes I_t = \Sigma_*. \tag{6.3.2}$$

That is, each element of $V_*(t)$ is expanded into a diagonal block matrix with elements whose subscripts t's are of the natural order, changing from 1 to T. A simple example illustrating the notation for the heteroscedastic disturbance variance matrix is as follows:

$$\begin{bmatrix} \sigma^2(1)\,I_1 & & & \\ & \sigma^2(2)\,I_2 & & \\ & & \ddots & \\ & & & \sigma^2(T)\,I_T \end{bmatrix} = \sigma^2(t) \otimes I_t. \tag{6.3.3}$$

By the use of the time-ordered Kronecker expansion, the following properties hold. If $a(t)$ and $b(t)$ are cross-section scalars, then

$$a(t)\,I_t \cdot b(t)\,I_t = a(t)\,b(t)\,I_t. \tag{6.3.4}$$

If $A(t)$ and $B(t)$ are cross-section matrices of the same size, then

$$A(t) \otimes I_t \cdot B(t) \otimes I_t = A(t)\,B(t) \otimes I_t. \tag{6.3.5}$$

The proofs of the above relations are straightforward by the direct multiplication of the expanded matrices. The most important and interesting relation is

$$\Sigma_*^{-1} = V_*^{-1}(t) \otimes I_t, \tag{6.3.6}$$

which may be proved by the direct multiplication of (6.3.6) and (6.3.2) with the application of (6.3.5) to yield an identity matrix.

To invert Σ_*, we only have to invert the cross-section matrix $V_*(t)$ and then use the time-ordered Kronecker expansion.

The determinant of $V_*(t)$ is

$$\det V_*(t) = \frac{\left[\prod\limits_{k=1}^{r-1} q_k(t)\right]\left[1 - \sum\limits_{k=1}^{r-1} q_k(t)\right]}{N(t)^{r-1}}, \tag{6.3.7}$$

which can be obtained by a few applications of elementary operations on (6.3.1). With a procedure similar to that of obtaining (6.3.7), the cofactors of $V_*(t)$, denoted as V_{*ij}, are

$$V_{*ij} = \frac{\prod\limits_{k=1}^{r-1} q_k(t)}{N(t)^{r-2}}, \qquad \text{when } i \neq j, \tag{6.3.8}$$

and

$$V_{*jj} = \frac{\prod\limits_{k \neq j}^{r-1} q_k(t)\left[1 - \sum\limits_{k \neq j}^{r-1} q_k(t)\right]}{N(t)^{r-2}}. \tag{6.3.9}$$

The elements and inverse of $V_*(t)$, denoted as V_*^{ij}, are

$$V_*^{ij} = \frac{N(t)}{q_r(t)} + \frac{N(t)}{q_j(t)}, \quad \text{for } i = j, \qquad (6.3.10)$$

$$V_*^{ij} = \frac{N(t)}{q_r(t)}, \quad \text{for } i \neq j \qquad (6.3.11)$$

and

$$V_*^{-1}(t) = \begin{bmatrix} \dfrac{N(t)}{q_r(t)} + \dfrac{N(t)}{q_1(t)} & \dfrac{N(t)}{q_r(t)} & \cdots & \dfrac{N(t)}{q_r(t)} \\[2ex] \dfrac{N(t)}{q_r(t)} & \dfrac{N(t)}{q_r(t)} + \dfrac{N(t)}{q_2(t)} & \cdots & \dfrac{N(t)}{q_r(t)} \\[2ex] \vdots & \vdots & \ddots & \vdots \\[2ex] \dfrac{N(t)}{q_r(t)} & \dfrac{N(t)}{q_r(t)} & \cdots & \dfrac{N(t)}{q_r(t)} + \dfrac{N(t)}{q_{r-1}(t)} \end{bmatrix}. \quad (6.3.12)$$

Thus, the inverse of Σ_* is a $((r-1)T \times (r-1)T)$ matrix with $(r-1)$ $(r-1)$ diagonal submatrices of size $(T \times T)^3$ with

$$\Sigma_*^{-1} = \begin{bmatrix} \Sigma^{11} & \Sigma^{12} & \cdots & \Sigma^{1,r-1} \\ \Sigma^{21} & \Sigma^{22} & \cdots & \Sigma^{2,r-1} \\ \vdots & \vdots & & \vdots \\ \Sigma^{r-1,1} & \Sigma^{r-2,2} & \cdots & \Sigma^{r-1,r-1} \end{bmatrix}, \qquad (6.3.13)$$

where

$$\Sigma^{ij} = \begin{bmatrix} \dfrac{N(1)}{q_r(1)} & & & \\ & \dfrac{N(2)}{q_r(2)} & & \\ & & \ddots & \\ & & & \dfrac{N(T)}{q_r(T)} \end{bmatrix}, \quad \text{for } i \neq j, \qquad (6.3.14)$$

---

[3] For a matrix similar to Σ_*^{-1}, see Zellner and Lee (1965) and for the matrix V_*^{-1}, see Kendall and Stuart (1958, p. 356).

and

$$\Sigma^{jj} = \begin{bmatrix} \dfrac{N(1)}{q_r(1)} + \dfrac{N(1)}{q_1(1)} & & & \\ & \dfrac{N(2)}{q_r(2)} + \dfrac{N(2)}{q_2(2)} & & \\ & & \ddots & \\ & & & \dfrac{N(T)}{q_r(T)} + \dfrac{N(T)}{q_{r-1}(T)} \end{bmatrix}. \quad (6.3.15)$$

6.4. Aitken's generalized unrestricted and restricted least squares estimators

Given the model (6.2.2) or (6.2.3) with assumptions (6.2.4) and (6.2.5), within the framework of Aitken's generalized least squares, the problem is to minimize

$$(y_* - X_* p_*)' \Sigma_*^{-1} (y_* - X_* p_*), \quad (6.4.1)$$

subject to

$$R p_* \leq \eta_r, \quad (6.4.2)$$

and

$$p_* \geq 0, \quad (6.4.3)$$

where η_r is a $(r \times 1)$ vector of ones and R, a submatrix of G, is a matrix $[I_1, I_2, ..., I_{r-1}]$ with each I a $(r \times r)$ identity matrix.

If there were no constraints (6.4.2) and (6.4.3), one might differentiate (6.4.1) with respect to p_* and set it equal to zero to obtain

$$-2X_*' \Sigma_*^{-1} (y_* - X_* \hat{p}_*) = 0, \quad (6.4.4)$$

which yields the unrestricted estimator

$$\hat{p}_* = (X_*' \Sigma_*^{-1} X_*)^{-1} X_*' \Sigma_*^{-1} y_*. \quad (6.4.5)$$

The covariance matrix of the estimator p_*, according to Aitken's generalized Gauss–Markov least squares theorem (1934), is then

$$V(p_*) = (X_*' \Sigma_*^{-1} X_*)^{-1}. \quad (6.4.6)$$

The deleted parameter p_r may be estimated as

$$\hat{p}_r = \eta_r - R\hat{p}_*. \tag{6.4.7}$$

We note that the estimate (6.4.5) is available if the covariance matrix of the disturbances Σ_* is known. In practice, the matrix Σ_* is usually unknown. Since the matrix Σ_* is a function of $q_j(t)$'s, a consistent way to estimate Σ_* is to replace the unknown true proportion $q_j(t)$'s by the known observed proportion $y_j(t)$'s in (6.3.13). As has been shown in § 3.2a, the $y_j(t)$'s are consistent estimates of the $q_j(t)$'s.

We also note that the estimate (6.4.5) may not fulfill the constraints (6.4.2) and (6.4.3), and as a consequence, the unrestricted weighted least squares estimates $\hat{p}_*$ may be negative or larger than unity.

To impose the restrictions (6.4.2) and (6.4.3) on the objective function (6.4.1), a quadratic programming algorithm may be used. Proceeding as in ch. 3, a simple procedure is to use the reducibility theorem of non-linear programming[4] to obtain the following equivalent linear programming problem: To maximize

$$(X_*'\Sigma_*^{-1}y_* - X_*'\Sigma_*^{-1}X_*\hat{p}_*^c)'\,p_*, \tag{6.4.8}$$

subject to

$$Rp_* \leq \eta_r \tag{6.4.2}$$

and

$$p_* \geq 0, \tag{6.4.3}$$

where $\hat{p}_*^c$ is the optimal solution we are seeking. The dual problem is then to minimize

$$\lambda'\eta_r, \tag{6.4.9}$$

subject to

$$R'X_* \geq X_*'\Sigma_*^{-1}y_* - X_*'\Sigma_*^{-1}X_*\hat{p}_*^c, \tag{6.4.10}$$

and

$$\lambda \geq 0, \tag{6.4.11}$$

where λ is a row vector of dual variables.

Given these linear formulations, let us now redefine $\hat{p}^c$ as p_* and define the following problem: To maximize

$$(X_*\Sigma_*^{-1}y_* - X_*'\Sigma_*^{-1}X_*p_*)'\,p_* - \lambda'\eta_r = -\lambda'p_r - \beta'p_* \leq 0 \tag{6.4.12}$$

---

[4] For a discussion of this theorem, see for example Charnes and Cooper (1961) and Kuhn and Tucker (1951).

subject to

$$Rp_* + p_r = \eta_r, \tag{6.4.13}$$

$$R'\lambda + (X'_* \Sigma_*^{-1} X_*) p_* - \beta = X'_* \Sigma_*^{-1} y_*, \tag{6.4.14}$$

and

$$p_*, p_r, \lambda, \beta \geq 0, \tag{6.4.15}$$

where p_r and β are vectors of slack variables. In this problem, p_r is also a subvector of p that has been deleted in forming (6.2.2). In the above formulation, the maximum value of the objective value is zero due to the equivalence of primal and dual problems. Thus, from the right-hand side of (6.4.12), we know that if the optimal solution $\hat{p}_*^c$ exists, then

$$\lambda^\tau \hat{p}_r^c + \beta^\tau \hat{p}_*^c = 0. \tag{6.4.16}$$

For eq. (6.4.16) to hold, either λ^τ or $\hat{p}_r^c$ or both and either β^τ or $\hat{p}_*^c$ or both must be zero because all of them cannot be negative. With this condition, the above problem is solvable by the use of the standard simplex method and the corresponding simplex tableau is listed in table 6.1.

TABLE 6.1

The simplex tableau for the restricted generalized
least squares estimates

β_0	$\lambda \geq 0$	$p_* \geq 0$	$p_r \geq 0$	$\beta \geq 0$
η_r	0	R	I	0
$X'_* \Sigma_*^{-1} y_*$	R'	$X'_* \Sigma_*^{-1} X_*$	0	$-I$

6.5. Results from the sampling experiment

In computing both unrestricted and restricted estimates, one meets the practical problem that for Σ_*^{-1} to exist, the $q_j(t)$'s or $y_j(t)$'s must not be zero. In order to avoid zero proportions, the aggregate *data from periods 3 to 14 are used* and some of the data are amended. That is, the $y_j(t)$'s which are zero are replaced by 0.0001. The mean and root square error statistics for the unrestricted estimates of 50 problems for each level of sample size 25, 50 and 100 are listed in table 6.2.

TABLE 6.2

Means and root square errors for the unrestricted generalized least squares estimates

Sample size	Means				Root mean square error			
25	0.4151	0.5778	0.1033	−0.0962	0.2743	0.3364	0.4506	0.3646
	0.1349	0.3788	0.4520	0.0343	0.1380	0.2459	0.3144	0.2361
	0.0312	0.1220	0.5747	0.2675	0.1094	0.1445	0.2235	0.1500
	0.0120	−0.0388	0.1978	0.8342	0.1323	0.0230	0.1993	0.1819
50	0.4754	0.5287	0.0057	−0.0087	0.1858	0.2635	0.2941	0.1981
	0.1390	0.3846	0.5188	−0.0424	0.1359	0.2213	0.2774	0.2007
	0.0149	0.1518	0.5540	0.2793	0.1174	0.1846	0.2160	0.1576
	−0.0088	−0.0457	0.2153	0.8932	0.0925	0.1418	0.1697	0.1238
100	0.5059	0.5511	−0.0347	−0.0221	0.1737	0.2762	0.3139	0.2519
	0.1727	0.3032	0.5310	−0.0068	0.1490	0.2634	0.3066	0.2604
	−0.0162	0.1982	0.5988	0.2193	0.0755	0.1387	0.1531	0.1064
	−0.0002	−0.0495	0.1598	0.8899	0.0496	0.0943	0.1083	0.0715

TABLE 6.3

Means and root square errors for the restricted generalized least squares estimates

Sample size	Means				Root mean square error			
25	0.4003	0.4875	0.1035	0.0088	0.2601	0.2113	0.1871	0.0224
	0.1370	0.4456	0.4061	0.0113	0.1007	0.1876	0.1588	0.0238
	0.0260	0.0743	0.6777	0.2219	0.0449	0.0801	0.1344	0.0937
	0.0125	0.0064	0.1083	0.8728	0.0257	0.0158	0.0991	0.0983
50	0.4787	0.4583	0.0592	0.0038	0.1736	0.1740	0.1131	0.0111
	0.1194	0.4781	0.3948	0.0077	0.0978	0.1555	0.1113	0.0209
	0.0112	0.0775	0.7106	0.2007	0.0269	0.0670	0.0932	0.0648
	0.0046	0.0096	0.0955	0.8722	0.0100	0.0242	0.0639	0.1299
100	0.5231	0.4197	0.0565	0.0007	0.1362	0.1483	0.1053	0.0034
	0.1261	0.4769	0.3914	0.0056	0.0824	0.1237	0.0951	0.0140
	0.0048	0.0962	0.7144	0.1846	0.0116	0.0411	0.0553	0.0399
	0.0012	0.0043	0.0849	0.9096	0.0038	0.0142	0.0514	0.0493

In table 6.2, the results reflect the property that the mean square error diminishes as the sample size increases. Since the aggregate data are not quite the same as the data used in previous chapters, direct comparisons cannot be made. However, some new estimates for the estimation methods given previously, which are based on the amended data, are also obtained and the comparison is made in ch. 12.

The mean and root mean square error statistics for the restricted estimates are listed in table 6.3. Again, the root mean square errors for the restricted estimates diminish when sample size increases. When a comparison is made between the unrestricted and the restricted estimators, the restricted estimator yields results which are closer to the true parameters (4.1.1). The sum of the root mean square errors of the restricted and unrestricted estimators are listed in table 6.4.

TABLE 6.4

Sum of the root mean square errors of the restricted and unrestricted generalized least squares estimates

Sample size	Unrestricted	Restricted
25	3.6036	1.7440
50	2.9800	1.3371
100	2.7396	0.9750

In this comparison, the restricted estimator is much better than the unrestricted estimator, especially when sample size is large. As the sample size increases, the ratio of the mean square errors between the unrestricted and restricted estimates increases from about 2 to 3.

The test of the goodness-of-fit of the sample results to normal distributions is also made for each element of the estimated matrix. The unrestricted estimates for size 25, 50 and 100 are all normally distributed in that the computed Kolmogorov–Smirnov D statistics are all smaller than the tabled value, 0.17 for 50 samples at the 10 percent significance level.

The restricted estimates are also normally distributed except for those elements whose target values are zero, such as p_{13}, p_{14}, p_{24}, p_{31}, p_{41} and p_{42}. For these parameters more than 50 percent of the estimates hit the target value. The percentages of the estimates hitting the target value for the three sample sizes are given in table 6.5.

Generally speaking, as sample size increases, the percentage of the estimates hitting the target value becomes higher.

TABLE 6.5
The percentages of hitting the target value
for the generalized least squares estimates

Elements	Sample sizes		
	25 (%)	50 (%)	100 (%)
p_{13}	54	56	54
p_{14}	76	82	92
p_{24}	64	80	76
p_{31}	42	70	72
p_{41}	54	72	88
p_{42}	80	70	88

CHAPTER 7

THE MINIMUM CHI-SQUARE ESTIMATOR

7.1. Preliminaries

Assume we have T time periods and in each time period that we have observations relating to $N(t)$ micro units. Each observation in the tth period is classified into one of r mutually exclusive and exhaustive classes (Markov states). If $q_j(t)$ is the probability of an observation (micro unit) in the tth time period falling into the jth class, we therefore have

$$\sum_j q_j(t) = 1. \qquad (7.1.1)$$

Thus, we have reduced the problem to one concerning a set of T multinomial distributions. If we let $n_j(t)$ be the number of micro units falling into the jth class at time period t, then

$$y_j(t) = n_j(t)/N(t), \qquad j = 1, 2, ..., r, \qquad (7.1.2)$$

are the corresponding relative frequencies (observed proportions). The probabilities $q_j(t)$ are assumed to follow a first order Markov process and are functions of a set of unknown constant parameters p_{ij} ($i, j = 1, 2, ..., r$) such that

$$q_j(t) = \sum_{i=1}^{r} y_i(t-1) p_{ij}. \qquad (7.1.3)$$

If the last time period's observations are $y_i(t-1)$, then the true proportions for the current period are $q_j(t), j = 1, 2, ..., r$, while the current observed proportions are $y_j(t), j = 1, 2, ..., r$. The difference between $q_j(t)$ and $y_j(t)$ is attributed to sampling error. To test the compatibility of the observed and theoretical frequencies, $N(t) y_j(t)$ and $N(t) q_j(t)$, the following chi-square statistic may be used:

$$\chi^2 = \sum_{t=1}^{T} \sum_{j=1}^{r} \frac{[N(t) y_j(t) - N(t) q_j(t)]^2}{[N(t) q_j(t)]}. \qquad (7.1.4)$$

The statistic (7.1.4) has the chi-square distribution with $T(r-1)$ degrees of freedom and the estimator obtained by minimizing the chi-square expression (7.1.4) is called a minimum chi-square estimator.

7.2. The restricted minimum chi-square estimator

To obtain the minimum chi-square (MCS) estimator, one may differentiate (7.1.4) with respect to p_{ij}.[1] Before differentiating (7.1.4) with respect to the p_{ij}'s, for minimizing purposes, it is necessary to express one of the $q_j(t)$'s, say $q_r(t)$, in terms of the remaining ones through the relations (7.1.1) and

$$\sum_{j=1}^{r} y_j(t) = 1. \tag{7.2.1}$$

If this is done, the expression in (7.1.4) assumes the form

$$\chi^2 = \sum_{t=1}^{T}\sum_{j=1}^{r-1} N(t)\,[(y_j(t) - q_j(t))^2/q_j(t)] + \sum_{t=1}^{T} N(t)\,[(y_r(t) - q_r(t))^2/q_r(t)]$$

$$= \sum_{t=1}^{T}\sum_{j=1}^{r-1} [N(t)/q_j(t)]\,(y_j(t) - q_j(t))^2 + \sum_{t=1}^{T} \Big([N(t)/q_r(t)]$$

$$\times \sum_{j=1}^{r-1} (y_j(t) - q_j(t))^2\Big)$$

$$= \sum_{t=1}^{T}\sum_{j=1}^{r-1} [N(t)/q_j(t)]\,(y_j(t) - q_j(t))^2 + \sum_{t=1}^{T} \Big([N(t)/q_r(t)]$$

$$\times \sum_{i=1}^{r-1}\sum_{j=1}^{r-1} (y_i(t) - q_i(t))(y_j(t) - q_j(t))\Big). \tag{7.2.2}$$

To simplify the equation (7.2.2), we let $\mathbf{y}'_* = [\mathbf{y}'_1, \mathbf{y}'_2, \cdots, \mathbf{y}'_{r-1}]$ with $\mathbf{y}'_j = [y_j(1), y_j(2), ..., y_j(T)]$, $(j = 1, 2, ..., r-1;\ t = 1, 2, ..., T)$, $X_*\boldsymbol{p}_* = \{q_j(t)\}$, $(j = 1, 2, ..., r-1;\ t = 1, 2, ..., T)$, D_* a $(T(r-1) \times T(r-1))$ diagonal matrix with elements $N(t)/q_j(t)$, $(j = 1, 2, ...,$

---

[1] Eq. (7.1.4) is a function of $q_j(t)$, which in turn is a function of p_{ij} by (7.1.3).

$r - 1$; $t = 1, 2, ..., T$) and S_* a $(T(r - 1) \times T(r - 1))$ matrix composed of $(r - 1)^2$ diagonal matrices of order $(T \times T)$ with all diagonal elements equal to $N(t)/q_r(t)$, where $t = 1, 2, ..., T$. Thus, the χ^2 expression (7.2.2) may be written as

$$\chi^2 = (y_* - X_* p_*)' D_* (y_* - X_* p_*) + (y_* - X_* p_*)' S_* (y_* - X_* p_*)$$

$$(7.2.3)$$

or more compactly, (7.2.3) is simply

$$\chi^2 = (y_* - X_* p_*)' \Sigma_*^{-1} (y_* - X_* p_*),$$ (7.2.4)

where Σ_*^{-1} is equal to the sum of D_* and S_* and specified in ch. 6 as (6.3.12). Thus, the expression for χ^2 is the same quadratic form encountered in the analysis of the model in ch. 6. Therefore, if we minimize (7.2.4) with respect to p_*, we obtain the unrestricted MCS estimator,[2]

$$\hat{p}_* = (X_*' \Sigma_*^{-1} X_*)^{-1} X_*' \Sigma_*^{-1} y_*,$$ (7.2.5)

which is the same as the unrestricted generalized least squares estimator. The restricted estimator may be obtained by minimizing (7.2.4) with respect to p_* subject to

$$R p_* < \eta_r$$ (7.2.6)

and

$$p_* \geq 0_r.$$ (7.2.7)

The result is the same as the generalized restricted least squares estimator, $\hat{p}_*^c$. Both estimators are subject to the condition that Σ_*^{-1} is known. In practice, an estimate, $\hat{\Sigma}_*^{-1}$, is usually used.

7.3. The modified restricted minimum chi-square estimator

To estimate the weight matrix Σ_*^{-1}, the true porportions $q_j(t)$'s are replaced by the observed proportions $y_j(t)$'s since the latter are consistent

---

[2] (7.2.5) is derived under the assumption that Σ_*'s elements are known and is therefore a conditional minimizing value. Since the elements of Σ_* involve the unknown parameters, it is necessary to employ iterative numerical techniques to obtain the unconditional minimizing value of p_*.

estimates of the former. If this is done, the chi-square expression becomes

$$(\chi')^2 = (y_* - X_* p_*)' \hat{\Sigma}_*^{-1} (y_* - X_* p_*), \qquad (7.3.1)$$

which, by the reverse process of rearrangement, as in (7.2.2), is equivalent to

$$(\chi')^2 = \sum_{t=1}^{T} \sum_{j=1}^{r-1} \frac{[N(t)\, y_j(t) - N(t)\, q_j(t)]^2}{[N(t)\, y_j(t)]}, \qquad (7.3.2)$$

and it is assumed that $y_j(t) \neq 0$. The expression (7.3.2) is called a modified chi-square and the resulting estimator is called the modified MCS estimator. Both the MCS and the modified MCS estimators have been shown to be best asymptotically normal (BAN) estimators (Neyman 1949).

7.4. An equivalent model

There is another way of looking at the problem of minimizing the chi-square expression (7.1.4). Previously, we noted that because of the relation (7.1.1), there are only $r - 1$ independent $q_j(t)$'s at each t. An alternative is to neglect the relation (7.1.1) temporarily and write (7.1.4) as

$$\chi^2 = \sum_{t=1}^{T} \sum_{j=1}^{r} [y_j(t) - q_j(t)]\, [N(t)/q_j(t)]\, [y_j(t) - q_j(t)]. \qquad (7.4.1)$$

Thus, if we let $y' = (y'_1, y'_2, ..., y'_r)$ with $y'_j = [y_j(1), y_j(2), ..., y_j(T)]$, Xp a vector of $q_j(t)$, $(j = 1, 2, ..., r; t = 1, 2, ..., T)$, and D a $(Tr \times Tr)$ diagonal matrix with elements $N(t)/q_j(t)$, $(j = 1, 2, ..., r; t = 1, 2, ..., T)$, then (7.4.1) is equal to

$$\chi^2 = (y - Xp)' D (y - Xp). \qquad (7.4.2)$$

By differentiating (7.4.2) with respect to p and assuming each p_{ij} independent,[3] we obtain the set of normal equations

$$X'DXp - X'Dy = 0. \qquad (7.4.3)$$

---

[3] This assumption clearly is at variance with what is assumed about the p_{ij} in the model. It is employed here to generate an 'unrestricted' estimator.

The unrestricted estimator is obtained by solving the equation (7.4.3) and is

$$\tilde{\tilde{p}} = (X'DX)^{-1} X'Dy. \tag{7.4.4}$$

The unrestricted estimator (7.4.4) may not be the same as the GLS unrestricted estimator,

$$\begin{bmatrix} \hat{p}_* \\ \hat{p}_r \end{bmatrix} = \begin{bmatrix} (X'_*\Sigma_*^{-1}X_*)^{-1} X'_*\Sigma_*^{-1}y_* \\ \eta_r - R\hat{p}_* \end{bmatrix}. \tag{7.4.5}$$

The restricted estimator from the alternate model (7.4.1) may be obtained by minimizing (7.4.1) subject to

$$Gp = \eta_r, \tag{7.4.6}$$

and

$$p \geq 0, \tag{7.4.7}$$

where G is an $(r \times r^2)$ matrix with r identity matrices in a row, i.e., $[I_1 \ I_2 \ \cdots \ I_r]$, where each I is an $(r \times r)$ identity matrix. As in previous chapters, by making use of the duality theorem and a primal–dual formulation, a quadratic programming problem results, which may be molded into the following simplex tableau (table 7.1). In table 7.1, λ_1

TABLE 7.1

The simplex tableau
for the MCS Estimator with independent
assumptions on parameters

p_0	λ_1	p	λ_2	β
η_r	0	G	0	0
$X'Dy$	G'	$X'DX$	$-G'$	$-I$

and λ_2 are the vectors of Lagrangean multipliers and β is the vector of slack variables. The restricted estimator from table 7.1 is the same as the estimator obtained from the procedure of § 7.3, or the GLS restricted estimator of ch. 6 since the same constraints are all finally imposed on the same objective function.

7.5. A numerical example

To illustrate application of the results in §§ 7.3 and 7.4 and to indicate the importance of taking account of restrictions, an absorbing chain problem with the following observations

$$[X_1, X_2] = \begin{bmatrix} 0.5000 & 0.5000 \\ 0.75 & 0.25 \\ 0.88 & 0.12 \\ 0.94 & 0.06 \\ 0.97 & 0.03 \end{bmatrix}, \qquad (7.5.1)$$

where $N(t) = N(t + 1) = 100$ will be used. Following the procedure of § 7.3, we can obtain an estimate of Σ_*^{-1} which for the present example is:

$$\hat{\Sigma}_*^{-1} = \begin{bmatrix} 533.33 \\ & 946.37 \\ & & 1773.00 \\ & & & 3436.40 \end{bmatrix}. \qquad (7.5.2)$$

Then,

$$X_*' \hat{\Sigma}_*^{-1} X_* = \begin{bmatrix} 5075.4790 & 691.9386 \\ 691.9386 & 230.4220 \end{bmatrix}, \qquad (7.5.3)$$

and

$$X_* \hat{\Sigma}_*^{-1} y_* = \begin{bmatrix} 5425.000 \\ 808.333 \end{bmatrix}. \qquad (7.5.4)$$

The inverse of $X_*' \hat{\Sigma}_*^{-1} X_*$ is

$$(X_*' \hat{\Sigma}_*^{-1} X_*)^{-1} = \begin{bmatrix} 0.0003 & -0.0010 \\ -0.0010 & 0.0073 \end{bmatrix}. \qquad (7.5.5)$$

Hence, the MCS estimator or the GLS estimator for p_1 is

$$\begin{bmatrix} \hat{p}_{11} \\ \hat{p}_{21} \end{bmatrix} = \begin{bmatrix} 1 \\ 0.5051 \end{bmatrix}. \qquad (7.5.6)$$

The deleted vector p_2 is then estimated as

$$\begin{bmatrix} \hat{p}_{12} \\ \hat{p}_{22} \end{bmatrix} = \begin{bmatrix} 0 \\ 0.4949 \end{bmatrix}. \qquad (7.5.7)$$

Therefore, the transition matrix is estimated to be

$$\hat{P} = \begin{bmatrix} 1 & 0 \\ 0.5051 & 0.4949 \end{bmatrix}. \tag{7.5.8}$$

The unrestricted estimates in this case satisfy the non-negativity constraints. Therefore, $\hat{P} = \hat{P}^c$.

Next, following the procedure of § 7.4, we obtain the weight matrix D as

$$D = \begin{bmatrix} 133\cdot33 & & & & & & & \\ & 113\cdot64 & & & & & & \\ & & 106\cdot38 & & & & & \\ & & & 103\cdot09 & & & & \\ & & & & 400\cdot00 & & & \\ & & & & & 833\cdot33 & & \\ & & & & & & 1666\cdot67 & \\ & & & & & & & 3333\cdot33 \end{bmatrix}. \tag{7.5.9}$$

The cross products weighted by D are

$$X'DX = \begin{bmatrix} 270.7295 & 71.6886 & & \\ 71.6886 & 42.3387 & & \\ & & 4804.7500 & 620.2500 \\ & & 620.2500 & 188.0833 \end{bmatrix}, \tag{7.5.10}$$

and

$$X'Dy = \begin{bmatrix} 307 \\ 93 \\ 307 \\ 93 \end{bmatrix}. \tag{7.5.11}$$

The inverse of $X'DX$ is

$$(X'DX)^{-1} = \begin{bmatrix} 0.0067 & -0.0013 & & \\ -0.0013 & 0.0428 & & \\ & & 0.0004 & -0.0012 \\ & & -0.0012 & 0.0093 \end{bmatrix}. \tag{7.5.12}$$

The unrestricted estimate of the transition matrix is then

$$\tilde{\tilde{P}} = [\tilde{\tilde{p}}_1 \tilde{\tilde{p}}_1] = \begin{bmatrix} 1.0012 & 0.0001 \\ 0.5013 & 0.4941 \end{bmatrix}. \tag{7.5.13}$$

The estimate $\tilde{p}_{11}$ violates the definition of probabilities in the sense that it is larger than unity. The restricted estimate based on table 7.1 with exactly 6 iterations is

$$\hat{P}^c = \begin{bmatrix} 1 & 0 \\ 0.5051 & 0.4949 \end{bmatrix},$$ (7.5.14)

which is the same as the GLS estimate. For computational simplicity, we recommend the use of the GLS procedure, since in the computational process, we eliminate a set of parameters.

CHAPTER 8

THE MACRO MAXIMUM LIKELIHOOD ESTIMATOR

As noted in a previous chapter, the observed proportions are considered to be generated from a set of multinomial distributions. Since the sample distribution is known, the maximum likelihood principle may be applied to obtain an estimator for the unknown parameters. As is well known, the minimum chi-square (MCS) estimator is obtained by minimizing the chi-square error expression while the maximum likelihood (ML) estimator is obtained by maximizing the likelihood function. According to the literature (Kendall and Stuart 1961, p.93), the MCS estimator and the ML estimator have the same large sample properties. Also both approaches may be considered as the primal and dual problems for finding a saddle point.

8.1. The multinomial distribution under the Lexis scheme

We assume that the observations $y_j(t)$, $(j = 1, 2, ..., r)$, the proportion of events which fall in the jth category in time period t, are generated by T sets of $N(t)$ independent trials each, with the probabilities of individuals falling in particular states constant within each set of trials, but varying from one set to another. This probability system can be regarded as a multinomial distribution under the assumptions of the Lexis scheme (Reitz 1934). That is, in the tth set, any one of the events s_1, s_2, ..., s_r can occur with respective probabilities $q_1(t), q_2(t), ..., q_r(t)$ on a single trial with

$$\sum_{j=1}^{r} q_j(t) = 1,$$

(8.1.1)

and

$$q_j(t) = \sum_{i=1}^{r} y_i (t - 1) p_{ij}.$$

(8.1.2)

Thus, if $N(t)$ trials are made at time t, the probability that s_1 occurs $n_1(t)$ times, s_2 occurs $n_2(t)$ times, etc. with

$$\sum_{i=1}^{r} n_i(t) = N(t), \tag{8.1.3}$$

is given by

$$f\left(n(t)|q(t), N(t)\right) = \frac{N(t)!}{\displaystyle\sum_{i=1}^{r} n_i(t)!} \sum_{j=1}^{r} q_j(t)^{n_j(t)}, \tag{8.1.4}$$

subject to (8.1.1), where $n(t)$ is a vector with elements $n_i(t)$ and $q(t)' = [q_1(t), q_2(t), ..., q_r(t)]$. The joint density function for outcomes for independent trials at time $t = 1, 2, ..., T$ is then

$$f(n|q, N) = \prod_{t=1}^{T} \frac{N(t)!}{\displaystyle\prod_{i=1}^{r} n_i(t)!} \prod_{j=1}^{r} q_j(t)^{n_j(t)}, \tag{8.1.5}$$

where $q' = [q(1)', q(2)', ..., q(T)']$, $N' = [N(1), N(2), ..., N(T)]$, $n = [n(1), ..., n(t)]$ and $q_j(t)$ is defined by (8.1.2).[1]

Since (8.1.1) and (8.1.3) hold, there are only $(r - 1)$ independent variables $n_j(t)$'s and $(r - 1)$ independent $q_j(t)$'s for each t. Let $n_r(t)$ and $q_r(t)$ be replaced by conditions (8.1.1) and (8.1.3). We may then write (8.1.5) in the form

$$f(n|q, N) = \prod_{t=1}^{T} \left[\frac{N(t)}{[\prod_i n_i(t)!]\,[N(t) - \sum_k n_k(t)]!} \prod_j q_j(t)^{n_j(t)} \times \right.$$
$$\left. \times \left(1 - \sum_k q_k(t)\right)^{N(t) - \Sigma_k n_k(t)} \right], \tag{8.1.6}$$

for $i, j, k = 1, 2, ..., r - 1$. Making use of (8.1.2) and defining $n_i(t)$ as

$$n_i(t) = N(t)\, y_i(t), \qquad i = 1, 2, ..., r, \tag{8.1.7}$$

the joint density function (8.1.6), viewed as a function of its transition probability parameters p_{ij}, given the sample data $N(t)$ and observed

---

[1] For a similar problem, see example 18.10 in Kendall and Stuart (1961, p. 50).

proportions $y_i(t)$, is the likelihood function

$$L(p|y) = \prod_{t=1}^{T} \frac{N(t)}{\prod_m (N(t)\, y_m(t))!\, \left(N(t) - \sum_k N(t)\, y_k\,(t-1)\right)!} \times$$

$$\times \prod_j \left(\sum_i y_i\,(t-1)\, p_{ij}\right)^{N(t)y_j(t)} \left(1 - \sum_k \sum_i y_i\,(t-1)\, p_{ik}\right)^{N(t) - \Sigma_k N(t) y_k(t)},$$

$$(8.1.8)$$

for $(m, j, k = 1, 2, ..., r - 1)$ and $(i = 1, 2, ..., r)$, where p denotes a column vector of parameters p_{ij}, $(i, j = 1, 2, ..., r)$ and y denotes a vector with elements $y_i(t)$ and $y_i\,(t-1)$, $(t = 1, 2, ..., T)$. Since the transition probabilities must satisfy certain conditions, the maximum likelihood (ML) estimator of the p_{ij}'s may be obtained by maximizing (8.1.8) subject to

$$\sum_j p_{ij} = 1, \qquad i = 1, 2, ..., r \qquad (8.1.9)$$

and

$$0 \le p_{ij} \le 1, \qquad \text{for all } i, j. \qquad (8.1.10)$$

Maximizing the likelihood function (8.1.8) without (8.1.9) and (8.1.10) may not yield ML estimates.

8.2. The mode of the likelihood function

As noted in the previous section, the mode of the likelihood function (8.1.8) may unfortunately fall outside the feasible region (8.1.10). Therefore, what we look for is a local maximum within the region (8.1.10) instead of the absolute maximum which may not be in the region (8.1.10). The former gives the restricted estimator and the latter gives the unrestricted estimator. Sometimes, the absolute maximum may fall in the feasible region and in this case, the local maximum is the global maximum, and the unrestricted estimator coincides with the restricted estimator.

Initially, let us ignore the conditions (8.1.9) and (8.1.10) on the p_{ij}. Following the usual procedures in deriving the ML estimator for the parameters of the likelihood function, we take the first derivatives of the logarithmic function, L, of (8.1.8) with respect to p_{ij} and set them equal to zero to obtain

$$\frac{\partial \log L}{\partial p_{ij}} = \sum_{t} \left[\frac{\partial \log L}{\partial q_j(t)} \cdot \frac{\partial q_j(t)}{\partial p_{ij}} \right] = 0, \qquad (8.2.1)$$

$$\text{for } j = 1, 2, ..., r - 1,$$

with

$$\frac{\partial \log L}{\partial q_j(t)} = \frac{n_j(t)}{q_j(t)} - \frac{N(t) - \sum\limits_{k} n_k(t)}{1 - \sum\limits_{k} q_k(t)}, \qquad (8.2.2)$$

$$\text{for } k = 1, 2, ..., r - 1,$$

and

$$\frac{\partial q_j(t)}{\partial p_{ij}} = \frac{\partial}{\partial p_{ij}} \sum_{i} y_i(t - 1) p_{ij} = y_i(t - 1), \qquad (8.2.3)$$

$$\text{for } i = 1, 2, ..., r.$$

Thus, the necessary maximizing conditions are

$$\sum_{t} \left[\frac{n_j(t)}{q_j(t)} - \frac{N(t) - \sum\limits_{k} n_k(t)}{1 - \sum\limits_{k} q_k(t)} \right] y_i(t - 1) = 0, \qquad (8.2.4)$$

$$\text{for } i = 1, 2, ..., r \quad \text{and} \quad j = 1, 2, ..., r - 1.$$

With rearrangement, equation (8.2.4) may be written as

$$\sum_{t=1}^{T} y_i(t - 1) \left(\frac{N(t)}{q_r(t)} + \frac{N(t)}{q_j(t)} \right) y_j(t) + \sum_{k \neq j}^{r-1} \sum_{t=1}^{T} y_i(t - 1) \frac{N(t)}{q_r(t)} y_k(t)$$

$$- \left[\sum_{t=1}^{T} y_i(t - 1) \left(\frac{N(t)}{q_r(t)} + \frac{N(t)}{q_j(t)} \right) \sum_{i=1}^{r} y_i(t - 1) p_{ij} \right.$$

$$\left. + \sum_{k \neq j}^{r-1} \sum_{t=1}^{T} y_i(t - 1) \frac{N(t)}{q_r(t)} \sum_{i=1}^{r} y_i(t - 1) p_{ik} \right] = 0, \qquad (8.2.5)$$

$$\text{for } i = 1, 2, ..., r \text{ and } j = 1, 2, ..., r - 1.$$

We may also write these $r(r - 1)$ equations in a compact matrix notation as

$$X'_* \Sigma_*^{-1} y_* - X'_* \Sigma_*^{-1} X_* p_* = 0, \qquad (8.2.6)$$

where p_* is a column vector of parameters with $(r-1)$ subvectors; X_* is a $(T(r-1) \times r(r-1))$ block diagonal matrix with

$$X_j = \begin{bmatrix} y_1(0) & y_2(0) & \cdots & y_r(0) \\ y_1(1) & y_2(1) & \cdots & y_r(1) \\ \vdots & \vdots & \ddots & \vdots \\ y_1(T-1) & y_2(T-1) & \cdots & y_r(T-1) \end{bmatrix}, \quad j = 1, 2, \ldots, r-1 \tag{8.2.7}$$

on the main diagonal; y_* is a $(T(r-1) \times 1)$ column vector with subvectors

$$y_j = \begin{bmatrix} y_j(1) \\ y_j(2) \\ \vdots \\ y_j(T) \end{bmatrix}, \quad \text{for } j = 1, 2, \ldots, r-1, \tag{8.2.8}$$

and Σ_*^{-1} is a $(T(r-1) \times T(r-1))$ matrix with $((r-1) \times (r-1))$ diagonal submatrices of size $(T \times T)$. That is,

$$\Sigma_*^{-1} = \begin{bmatrix} \Sigma^{11} & \Sigma^{12} & \cdots & \Sigma^{1,r-1} \\ \vdots & & \ddots & \vdots \\ \Sigma^{r-1,1} & & \cdots & \Sigma^{r-1,r-1} \end{bmatrix}, \tag{8.2.9}$$

where

$$\Sigma^{ij} = \begin{bmatrix} \dfrac{N(1)}{q_r(1)} & & & \\ & \dfrac{N(2)}{q_r(2)} & & \\ & & \ddots & \\ & & & \dfrac{N(T)}{q_r(T)} \end{bmatrix}, \quad \text{for } i \neq j, \tag{8.2.10}$$

and

$$\Sigma^{jj} = \begin{bmatrix} \dfrac{N(1)}{q_j(1)} + \dfrac{N(1)}{q_r(1)} & & & \\ & \dfrac{N(2)}{q_j(2)} + \dfrac{N(2)}{q_r(2)} & & \\ & & \ddots & \\ & & & \dfrac{N(T)}{q_j(T)} + \dfrac{N(T)}{q_r(T)} \end{bmatrix}. \tag{8.2.11}$$

The matrix Σ_* is the variance–covariance matrix of the $y_j(t)$'s, $j = 1, 2, ..., r - 1$, (Kendall and Stuart 1958, pp. 355–356). That is, Σ_* is a block diagonal matrix with $(r - 1)^2$ submatrices Σ_{ij} such that

$\Sigma_{ij} = $ a $(T \times T)$ diagonal matrix with elements

$$-[q_i(t)\, q_j(t)]/N(t), \quad \text{for } t = 1, 2, ..., T \text{ and } i \neq j;$$

$\Sigma_{ii} = $ a $(T \times T)$ diagonal matrix with elements

$$[q_i(t)\, (1 - q_i(t))]/N(t), \quad \text{for } t = 1, 2, ..., T.$$

Thus, if the covariance matrix Σ_* is known, then the unrestricted ML estimator is

$$\hat{p}_* = (X'_* \Sigma_*^{-1} X_*)^{-1} X'_* \Sigma_*^{-1} y_*. \tag{8.2.12a}$$

The deleted parameter vector p_r may be estimated from the relation

$$\hat{p}_r = \eta_r - R\hat{p}_*, \tag{8.2.12b}$$

where R is an $(r \times (r - 1))$ known coefficient matrix $[I_1, I_2, ..., I_{r-1}]$, with each I_i an $(r \times r)$ identity matrix and η_r is an $(r \times 1)$ column vector with all entries unity. The solution set $(\hat{p}_*, \hat{p}_r)$ is unique and independent of which column of P is designated as p_r, i.e., invariant under permutation of the states. The estimator (8.2.12) is the same as the generalized least squares estimator presented in ch. 6.

The covariance matrix of the estimates (8.2.12) may be obtained by taking the inverse of the information matrix[2] which is defined according to Fisher (1961) as

$$Ip_* = -E\, [\partial^2 \log L/\partial p_{ij}\, \partial p_{kh}]. \tag{8.2.13}$$

To obtain (8.2.13), we differentiate the left-hand side of (8.2.5) and then take the expectation with negative sign to yield

$$-E\, \frac{\partial^2 \log L}{\partial p_{ij}\, \partial p_{kh}}$$

$$= \begin{cases} \displaystyle\sum_{t=1}^{T} q_i\,(t-1)\left(\frac{N(t)}{q_j(t)} + \frac{N(t)}{q_r(t)}\right) q_k\,(t-1)\left(\frac{N(t-1)-1}{N(t-1)}\right) & \text{for } j = h, \\[3ex] \displaystyle\sum_{t=1}^{T} q_i\,(t-1)\left(\frac{N(t)}{q_r(t)}\right) q_k\,(t-1)\left(\frac{N(t-1)-1}{N(t-1)}\right) & \text{for } j \neq h, \end{cases}$$

$$\tag{8.2.14}$$

---

[2] For an example, see Zellner and Tiao (1964, p. 771).

in which the following expectations are utilized:

$$E(y_i(t)) = q_i(t),$$

$$E(y_i(t) y_j(t)) = [(N(t) - 1)/N(t)] q_i(t) q_j(t),$$

$$E(y_i(t) y_j(t-1)) = q_i(t) q_j(t-1)$$

and

$$E(y_i(t) y_j(t-1) y_k(t-1))$$
$$= [(N(t-1) - 1)/N(t-1)] q_i(t) q_j(t-1) q_k(t-1), \quad \text{etc.}$$

The analysis goes forward given the $N(t)$ and when sample size at each time t is large, the finite adjustment factor $[(N(t-1)-1)/N(t-1)]$ is approximately 1 and $q_i(t-1)$, $q_k(t-1)$ and also $q_r(t)$ and $q_j(t)$ may be estimated by their observed proportions $y_i(t-1)$, etc. Thus, the estimated covariance matrix of p_* is approximately

$$V(\hat{p}_*) = (X'_* \Sigma_*^{-1} X_*)^{-1}. \tag{8.2.15}$$

The covariance matrix of all parameters $V(p)$ may be obtained by repermuting the Markov states and repeating the above procedures. It is interesting to note that the augmented matrix $V(\hat{p})$ is singular.

Because of the peculiar structure of X_* and Σ_*, the matrix $V(\hat{p})$ is symmetric and has row-sums or column-sums all zeros. The covariance matrix for the Telser (1963) problem is computed as an example and is listed in the following table (table 8.1).

8.3. The macro maximum likelihood estimator

The estimator given in (8.2.12) is an unrestricted estimator, since the potential restrictions defined by (8.1.9) and (8.1.10) have not been imposed. To insure the requirements placed on the transition probabilities are satisfied, a restricted ML estimator is necessary. Thus, to restate the original problem, we maximize the logarithm of the likelihood function (8.1.8) subject to the constraints (8.1.9) and (8.1.10) or equivalently,

$$Rp_* \leq \eta_r \tag{8.3.1}$$

and

$$p_* \geq 0. \tag{8.3.2}$$

TABLE 8.1

Covariances of the unrestricted ML estimates for the Telser example

	p_{11}	p_{21}	p_{31}	p_{12}	p_{22}	p_{31}	p_{31}	p_{23}	p_{33}
p_{11}	0.0495								
p_{21}	0.0086	0.0147							
p_{31}	-0.0746	-0.0280	0.1308						
p_{12}	-0.0239	-0.0056	0.0375	0.0438					
p_{22}	-0.0056	-0.0082	0.0164	0.0103	0.0151				
p_{32}	0.0375	0.0164	-0.0682	-0.0692	-0.0300	0.1256			
p_{13}	-0.0257	-0.0031	0.0370	-0.0799	-0.0047	0.0317	0.0456		
p_{23}	-0.0030	-0.0065	0.0116	-0.0048	-0.0068	0.0137	0.0078	0.0133	
p_{33}	0.0370	0.0116	-0.0626	0.0316	0.0316	-0.0574	-0.0687	-0.0242	0.1200

Since the likelihood function (8.1.8) is a non-linear function and the constraints (8.3.1) and (8.3.2) are linear, we may, as in ch. 3, use the reducibility theorem of non-linear programming to obtain the solution for the restricted problem. This formulation results in the following linear programming problem:
To maximize

$$(X_* \Sigma_*^{-1} y_* - X_*' \Sigma_*^{-1} X_* \hat{p}^c)' p_* \tag{8.3.3}$$

subject to (8.3.1) and (8.3.2). The constant part of (8.3.3) is from (8.2.6), evaluated at its unknown optimal solution $\hat{p}^c$, i.e., the tangent of the likelihood function at its local maximum. By making use of the duality theorem of Dorn (1960), the following dual problem results:
To minimize

$$\lambda' \eta_r \tag{8.3.4}$$

subject to

$$R'\lambda \geq (X_* \Sigma_*^{-1} y_* - X_*' \Sigma_*^{-1} X_* \hat{p}_*^c) \tag{8.3.5}$$

and

$$\lambda \geq 0, \tag{8.3.6}$$

where λ is a column vector of dual variables. Since both the primal and dual problems involve the unknown solution vector $\hat{p}_*^c$, we may formulate the following primal–dual programming problem:
To maximize

$$(X_*' \Sigma_*^{-1} y_* - X_*' \Sigma_*^{-1} X_* p_*)' p_* - \lambda' \eta_r = -\alpha' p_* - \lambda' p_r \leq 0 \tag{8.3.7}$$

subject to

$$Rp_* + p_r = \eta_r, \tag{8.3.8}$$

$$R'\lambda + (X_*' \Sigma_*^{-1} X_*) p_* - \alpha = X_*' \Sigma_*^{-1} y_* \tag{8.3.9}$$

and

$$p_*, p_r, \lambda, \alpha \geq 0, \tag{8.3.10}$$

where p_r and α are vectors of slack variables. This is the same quadratic programming problem as that of the restricted generalized least squares and the simplex tableau is the same as table 6.1 of ch. 6.

In obtaining the preceding estimates of the p_{ij}'s, we have assumed that Σ_*^{-1} is known. In practice, Σ_*^{-1} is unknown and is a function of the unknown p_{ij}. This uncomfortable aspect can be removed by an iterative feed back procedure to improve the estimates p_{ij}.[3] That is, initially the unknown element $q_j(t)$ in the matrix Σ_*^{-1} is replaced by its consistent estimate $y_j(t)$. The problem is then solved to yield a set of estimates denoted $\hat{P}^c(1) = \{\hat{p}_{ij}^c(1)\}$. Then

$$\hat{q}_j^c(t) = \sum_{i=1}^{r} y_i(t-1)\hat{p}_{ij}^c(1),$$

and the new solution $\hat{p}_{ij}^c(2)$ is obtained using $\hat{q}_j^c(t)$ in the estimation of Σ_*^{-1}. The feed back procedure is repeated until

$$\hat{p}_{ij}^c(n+1) = \hat{p}_{ij}^c(n).$$

Thus, a recursive quadratic programming problem results and $p_{ij}(n)$ or $p_{ij}(n+1)$ will be the exact estimate.[4]

8.4. Results from the sampling experiment

Since the ML estimator is exactly the same as the GLS estimator, as well as the MCS estimator, the sampling experiment results are the same as those listed in ch. 6.

8.5. Some applications

a. Results for the brand change problem

In contrast to the original assumption that the data were generated from a fixed sample of smokers, let us now assume that the number of cigarette smokers is proportional to the number of cigarettes sold. Assuming an average per capita consumption of one pack per day, the number of cigarette smokers for each year was computed and used as factors in determining the weight matrix for the maximum likelihood estimator. Under this specification, the initial solution for the maxi-

---

[3] For the successive approximation to ML estimators, see Kendall and Stuart (1961, pp. 48–51).

[4] The condition $0 \leq p_{ij} \leq 1$ insures that the recursive system will not explode.

mum likelihood estimator is

$$\hat{P}^c = [\hat{p}^c_{ij}] = \begin{bmatrix} 0.6774 & 0.1294 & 0.1933 \\ 0.0000 & 0.8828 & 0.1172 \\ 0.3935 & 0.0000 & 0.6065 \end{bmatrix}, \qquad (8.5.1)$$

which is quite close to the weighted restricted least squares result (5.6.1).

We noted in the formulation of the maximum likelihood estimator that a more accurate estimator may obtained by a feed back recursive computation procedure. If this procedure is applied to the Telser problem, the successive approximations are shown in table 8.2.

TABLE 8.2

Maximum likelihood estimator of the transition matrix of the cigarette consumers by recursive quadratic programming for the three leading brands: Camels, Lucky Strike and Chesterfield

Stage	Estimates			Recursive difference
Initial solution	0.6776	0.1290	0.1934	—
	0	0.8828	0.1172	
	0.3935	0	0.6065	
1st recursive solution	0.6699	0.1421	0.1880	0.0628
	0	0.8710	0.1290	
	0.4000	0	0.6000	
2nd recursive solution	0.6689	0.1424	0.1887	0.0049
	0	0.8707	0.1293	
	0.4012	0	0.5988	
3rd recursive solution	0.6689	0.1425	0.1886	0.0004
	0	0.8707	0.1293	
	0.4011	0	0.5989	

From table 8.2, we note that the 4th recursive solution is the same as the 3rd solution to 4 significant digits. Thus, the last solution is the exact maximum likelihood estimator. In this case, it does not deviate significantly from the initial solution. The convergence of the recursive solution is noted by the recursive error which is computed as the sum of the absolute deviations of the estimates in the new solution from those in the preceding solution. In this case, the recursive error converges to zero in a well-behaved manner.

TABLE 8.3

Number of farm families and percentages of classified farmers, Taiwan, 1941–1966

Year	Total number of farm families	Percentages		
		Owner	Part-owner	Tenant
1941	440,105	31	31	38
42	452,462	31	31	38
43	470,374	31	30	39
44	482,776	31	30	39
45	500,533	30	29	41
46	527,016	33	28	39
47	553,308	32	27	41
48	597,333	35	26	39
49	620,875	36	25	39
50	638,062	36	26	38
51	661,125	38	25	37
52	676,750	39	26	35
1953	702,325	55	24	21
54	716,582	57	24	19
55	732,555	59	24	17
56	746,318	60	23	17
57	759,234	60	23	17
58	769,925	61	23	16
59	780,402	62	23	15
60	785,592	64	21	15
61	800,835	65	21	14
62	809,917	65	21	14
63	824,560	66	21	13
64	834,827	66	21	13
65	847,242	67	20	13
66	854,203	67	21	12

Source: *Taiwan Agricultural Yearbook*, Department of Agriculture and Forestry, Provincial Government of Taiwan, 1941–1967.

b. Results for the tenure status problem

In addition to the Telser smoker behavior problem, let us now consider the problem of the tenure behavior pattern before and after the land reform program in Taiwan. That is, we would like to study, within the Markov framework, the behavior systems reflecting the time ordered changes in the tenure status of farmers before and after the new land policy. The land reform program which started in 1949 actually effectively got under way in 1953.

TABLE 8.4

Maximum likelihood estimators of transition probability matrix for the periods before and after the land reform program in Taiwan

Markov states	1941–1952			1953–1966		
	Tenant	Part-Owner	Owner	Tenant	Part-Owner	Owner
Tenant	0.86	0.06	0.08	0.66	0.34	0
Part-owner	0.18	0.82	0	0.21	0.54	0.25
Owner	0	0.07	0.93	0	0.07	0.93

Although time ordered tenure status data are not available, annual proportional data for the island's tenant, part owner and owner-operator are available from 1900 to 1966. For this illustration, we assume the aggregate data were generated by a first order Markov process and that the transition probability system was stationary for the years 1941–52. After 1952, we assume, due to the land reform, the transition probabilities changed from the previous period but were constant over the sample of years 1953–66. The proportion data for these years are given in table 8.3.

Using these data, the maximum likelihood transition probability estimates for the two periods are given in table 8.4.

The results appear to be consistent with the expected economic outcome and would suggest that the land reform program has provided tenants with a good opportunity of changing their tenant status.

CHAPTER 9

BAYESIAN ANALYSIS OF THE 'MACRO' MODEL

In ch. 2, a multivariate beta prior distribution was combined with the "micro" likelihood function by means of Bayes' theorem to obtain the posterior distribution for the transition probabilities. Similarly, in the present chapter, a multivariate beta prior distribution is combined with the likelihood function, based on the "macro" data, to obtain a posterior distribution for the transition probabilities. This posterior distribution can be employed to make inferences about the transition probability parameters. In particular, a Bayesian estimator for the transition probabilities will be developed in what follows.

9.1. Bayesian analysis: prior distribution

In this study, the following basic multivariate beta PDF (Mauldon 1961; Martin 1967) for the elements of the ith row of the transition probability matrix will be employed:

$$f(\boldsymbol{p}_i|\boldsymbol{a}_i) = \frac{\Gamma(\boldsymbol{\eta}'\boldsymbol{a}_i)}{\prod\limits_{j=1}^{r} \Gamma(a_{ij})} \prod_{j=1}^{r} p_{ij}^{a_{ij}-1}, \qquad (9.1.1)$$

with $\Sigma_{j=1}^{r} p_{ij} = 1$ and $0 \leqq p_{ij}, j = 1, 2, ..., r$, where

$\boldsymbol{p}_i' = (p_{i1}, p_{i2}, ..., p_{ij})$, a vector of transition probabilities,
$\boldsymbol{\eta}' = (1, 1, ..., 1)$, a $(1 \times r)$ vector of ones,
$\boldsymbol{a}_i' = (a_{i1}, a_{i2}, ..., a_{ir})$, a vector of prior parameters, with $a_{ij} > 0$, and
Γ denotes the gamma function.

From (9.1.1), the marginal distribution of a particular transition probability, say p_{ij}, is given by the following beta distribution:[1]

$$f(p_{ij}|\boldsymbol{a}_i) = \frac{1}{B\left(a_{ij}, \Sigma_{k \neq j}^r a_{ik}\right)} p_{ij}^{a_{ij}-1}(1 - p_{ij})^{\Sigma_{k \neq j}^r a_{ik}-1}, \quad (9.1.2)$$

with $0 \leq p_{ij} \leq 1$, where B denotes the beta function. As is well known, the mean and variance of the prior distribution in (9.1.2) are given by:

$$E(p_{ij}) = a_{ij} \bigg/ \sum_{j=1}^r a_{ij} = a_{ij}/a_i \quad (9.1.3)$$

and

$$\text{Var}(p_{ij}) = \frac{a_{ij}(a_i - a_{ij})}{a_i^2(a_i + 1)} = \frac{E(p_{ij})(1 - E(p_{ij}))}{a_i + 1}, \quad (9.1.4)$$

where $a_i = \Sigma_{j=1}^r a_{ij}$ and the a_{ij}'s are positive parameters of the prior PDF in (9.1.2). Further, the prior covariance between any two elements in the ith row of the transition probability matrix, say p_{ik} and p_{ih}, which can be obtained from the marginal bivariate prior PDF for p_{ik} and p_{ih}, is

$$\text{Cov}(p_{ik}, p_{ih}) = -\frac{a_{ik}a_{ih}}{a_i^2(a_i + 1)} = -\frac{Ep_{ik}Ep_{ih}}{a_i + 1}. \quad (9.1.5)$$

If the r rows of the transition probability matrix are assumed a priori to be independently distributed, each with a prior PDF in the form of (9.1.1), then the joint prior PDF for all the transition probabilities in the matrix $P = \{p_{ij}\}$ is given by the product of r factors each in the form of (9.1.1). That is, the joint prior PDF is:

$$f(P|A) = \prod_{i=1}^r f(\boldsymbol{p}_i|\boldsymbol{a}_i) = \prod_{i=1}^r \frac{\Gamma(a_i)}{\prod_{j=1}^r \Gamma(a_{ij})} \prod_{j=1}^r p_{ij}^{a_{ij}-1}, \quad (9.1.6)$$

where $A = \{a_{ij}\}$, an $(r \times r)$ matrix of prior parameters, $\boldsymbol{\eta}'\boldsymbol{p}_i = 1$ for $i = 1, 2, \ldots, r$, and $0 \leq p_{ij} \leq 1$. The joint prior PDF in (9.1.6) is in the matrix beta form (Martin 1967, pp. 141–146).

---

[1] To obtain (9.1.2) from (9.1.1), the following formulas are repeatedly used:

$$B(m, n) = [1/(a^{m+n-1})] \int_0^a y^{m-1}(a - y)^{n-1} \, \mathrm{d}y$$

and

$$B(m, n) = B(n, m).$$

The prior distribution in (9.1.6) will be combined below with the likelihood function to provide a posterior PDF for the elements of p. It is thought that (9.1.6) is flexible enough to represent prior information in a number of applications.

9.2. Likelihood function and posterior distribution

We assume that our data are in the form of aggregate proportions giving the proportion of units in each state in each of T time periods. As explained in ch. 8, these data are considered to be generated by a multinomial process under the Lexis scheme and under this assumption the likelihood function is:[2]

$$
L(\boldsymbol{p}_*|\boldsymbol{y}) = \prod_{t=1}^{T} \frac{N(t)!}{\prod_m [N(t)\, y_m(t)]!\, \left[N(t) - \sum_k N(t)\, y_k(t)\right]!} \times
$$
$$
\times \prod_j \left[\sum_i y_i(t-1) p_{ij}\right]^{N(t)\, y_j(t)} \left[1 - \sum_k \sum_i y_i(t-1) p_{ik}\right]^{N(t)[1 - \Sigma_k y_k(t)]},
$$
(9.2.1)

with $j, k, m = 1, 2, \ldots, r-1$ and $i = 1, 2, \ldots, r$. Further, in (9.2.1), $\boldsymbol{p}'_* = (p'_1, p'_2, \ldots, p'_{r-1})$, a vector of transition probabilities with p_j an $(r \times 1)$ vector and $\boldsymbol{y}' = [\boldsymbol{y}(0)', \boldsymbol{y}(1)', \ldots, \boldsymbol{y}(T)']$ is the vector of observed proportions. Note that we take the initial observation vector, $\boldsymbol{y}(0)$, as given.

On combining the prior PDF in (9.1.6) with the likelihood function in (9.2.1), the posterior PDF for the transition probabilities is

$$
f(\boldsymbol{p}_*|\boldsymbol{n}) = K\left[\prod_{t=1}^{T}\prod_{j=1}^{r-1} q_j(t)^{n_j(t)}\left[1 - \sum_{m=1}^{r-1} q_m(t)\right]^{N(t) - \Sigma_{m=1}^{r-1} n_m(t)} \times \right.
$$
$$
\left. \times \prod_{i=1}^{r}\prod_{j=1}^{r-1} p_{ij}^{a_{ij}-1}\left[1 - \sum_{k=1}^{r-1} p_{ik}\right]^{a_i - (\Sigma_{k=1}^{r-1} a_{ik} - 1)}\right],
$$
(9.2.2)

where K is a normalizing constant which does not depend on the transition probabilities, $n_j(t) = N(t)\, y_j(t)$, $\boldsymbol{n}' = [\boldsymbol{n}(0)', \boldsymbol{n}(1)', \ldots, \boldsymbol{n}(T)']$, and $q_j(t) = \Sigma_{i=1}^{r} y_i(t-1) p_{ij}$. In what follows, we shall analyze several properties of the posterior PDF in (9.2.2).

---

[2] For more details regarding the following likelihood function, see ch. 8.

9.3. The mode of the posterior distribution

In this section, we shall determine the value of p_* associated with the mode of the posterior distribution in (9.2.2). The modal value is, of course, a measure of location associated with the greatest posterior density. Given that posterior probabilities are regarded as measuring degrees of rational belief, the modal value is then the value of p_* such that the probability that p_* is in a small neighborhood containing the modal value, is greater than that associated with a small neighborhood containing another value of p_*. Of course, the modal value of the posterior distribution must be in the following acceptable region of the parameter space:

$$\sum_{j=1}^{r} p_{ij} = 1, \qquad i = 1, 2, ..., r \tag{9.3.1}$$

and

$$0 \le p_{ij} \le 1, \qquad \text{for all } i \text{ and } j. \tag{9.3.2}$$

Initially, let us ignore condition (9.3.2) in finding the modal value of (9.2.2). If the solution to this problem happens to satisfy (9.3.2), then we have a point estimate in the acceptable region of the parameter space. If (9.3.2) is not satisfied, then we shall have to resort to the programming approach presented in a later section of this chapter. On taking the logarithm of (9.2.2), differentiating with respect to the p_{ij}'s and setting these derivatives equal to zero, we obtain:

$$\sum_{t=1}^{T} \left[\frac{n_j(t)}{q_j(t)} - \frac{N(t) - \sum_{m=1}^{r-1} n_m(t)}{1 - \sum_{m=1}^{r-1} q_m(t)} \right] y_i(t-1) + \frac{a_{ij} - 1}{p_{ij}}$$

$$- \frac{a_i - \left(\sum_{k=1}^{r-1} a_{ik} - 1 \right)}{1 - \sum_{k=1}^{r-1} p_{ik}} = 0, \tag{9.3.3}$$

with $i = 1, 2, ..., r$ and $j = 1, 2, ..., r - 1$. Solving these $r(r-1)$ equations for the values of the $r(r-1)$ values of the p_{ij}'s satisfying (9.3.3) appears to be a difficult task. However, note that the $r(r-1)$

conditions in (9.3.3) may be rearranged as follows:

$$\sum_{t=1}^{T} y_i (t-1) \left[\frac{N(t)}{q_r(t)} + \frac{N(t)}{q_j(t)} \right] y_i(t) + \sum_{t=1}^{T} \sum_{k \neq j}^{r-1} y_i (t-1) \left[\frac{N(t)}{q_r(t)} \right] y_k(t)$$

$$- \sum_{t=1}^{T} y_i (t-1) \left[\frac{N(t)}{q_r(t)} + \frac{N(t)}{q_j(t)} \right] \sum_{i=1}^{r} y_i (t-1) p_{ij}$$

$$- \sum_{t=1}^{T} \sum_{k \neq 1}^{r-1} y_i (t-1) \left[\frac{N(t)}{q_r(t)} \right] \sum_{i=1}^{r-1} y_i (t-1) p_{ik}$$

$$+ \left[\frac{a_i + 1}{p_{ir}} + \frac{a_i + 1}{p_{ij}} \right] \left[\frac{a_{ij} - 1}{a_i + 1} \right] + \sum_{k \neq j}^{r-1} \left[\frac{a_i + 1}{p_{ir}} \right] \left[\frac{a_{ik} - 1}{a_i + 1} \right]$$

$$- \left[\frac{a_i + 1}{p_{ir}} + \frac{a_i + 1}{p_{ij}} \right] \left[\frac{a_i - r}{a_i + 1} \right] p_{ij}$$

$$- \sum_{k \neq j}^{r-1} \left[\frac{a_i + 1}{p_{ir}} \right] \left[\frac{a_i - r}{a_i + 1} \right] p_{ik} = 0 \qquad (9.3.4)$$

with $i = 1, 2, ..., r$ and $j = 1, 2, ..., r - 1$. In matrix terms, (9.3.4) may be rewritten as

$$X'_* \Sigma_*^{-1} y_* - X'_* \Sigma_*^{-1} X_* p_* + \Sigma_0^{-1} f - \Sigma_0^{-1} J p_* = 0, \qquad (9.3.5)$$

where Σ_0^{-1} is an $(r(r-1) \times r(r-1))$ block diagonal matrix with $(r-1)^2$ diagonal submatrices, Σ_0^{kj}, $k, j = 1, 2, ..., r - 1$. Each diagonal submatrix, Σ_0^{jj}, $j = 1, 2, ..., r - 1$, is of size $(r \times r)$ and has elements $(a_i + 1)(1/p_{ir} + 1/p_{ij})$, $i = 1, 2, ..., r$, on the diagonal. Each off diagonal submatrix, Σ_0^{kj}, $k \neq j$, is also of size $(r \times r)$ with elements $(a_i + 1)/p_{ir}$, $i = 1, 2, ..., r$, on the diagonal.[3] J is an $(r(r-1) \times r(r-1))$ diagonal matrix with elements $(a_i - r)/(a_i + 1)$, $i = 1, 2, ..., r; j = 1, 2, ..., r - 1$ (i.e., we have $r - 1$ identical diagonal blocks)

---

[3] A stronger assumption about the prior distribution is to assume, subjectively, each p_{ij} is an independent univariate beta distribution. Thus, the joint PDF of $r(r-1)$ independent parameters $(p_{ij}, i = 1, 2, ..., r; j = 1, 2, ..., r - 1)$ is

$$f(p_*) = \prod_{i,j} [1/B(a_{ij}, b_{ij})] p_{ij}^{a_{ij}-1} (1 - p_{ij})^{b_{ij}-1}.$$

This prior is a special case of (9.1.6) with (9.1.5) being zero. The resulting Bayesian estimator is similar to (9.3.5) except that Σ_0^{-1} becomes a diagonal matrix, since the covariances vanish.

and f is an $(r\,(r-1)\times1)$ column vector with elements $(a_{ij}-1)/(a_i+1)$ $i = 1, 2, ..., r$ and $j = 1, 2, ..., r-1$.

Since the matrices Σ_*^{-1} and Σ_0^{-1} in (9.3.5) have elements involving the unknown parameters, the p_{ij}'s, we shall obtain the modal value of the posterior distribution in (9.2.2) as follows. We substitute the prior means for the p_{ij}'s appearing in Σ_*^{-1} and Σ_0^{-1} and denote these expressions by $\hat{\Sigma}_*^{-1}$ and $\hat{\Sigma}_0^{-1}$ respectively. Then (9.3.5) can be solved to yield

$$\tilde{p}_*(1) = (X_*'\hat{\Sigma}_*^{-1}X_* + \hat{\Sigma}_0^{-1}J)^{-1}(X_*'\hat{\Sigma}_*^{-1}y_* + \hat{\Sigma}_0^{-1}f). \quad (9.3.6a)$$

This approximation to the modal value of the posterior distribution can be improved by substituting the elements of $\tilde{p}_*(1)$ into the expressions for Σ_*^{-1} and Σ_0^{-1} and then resolving (9.3.5) to obtain a new approximation to the modal value, say $\tilde{p}_*(2)$. This iterative procedure can be repeated until the results stabilize at a value which we denote by $\tilde{p}_*$, which will be the modal value of the posterior distribution. The modal value for the deleted parameter vector, p_r, is given by

$$\tilde{p}_r = \eta_r - R\tilde{p}_*. \quad (9.3.6b)$$

As will be shown below, an approximation to the variance–covariance matrix of the posterior distribution in (9.2.2) is given by

$$V(p_*) = (X_*'\hat{\Sigma}_*^{-1}X_* + \hat{\Sigma}_0^{-1}J)^{-1}, \quad (9.3.7a)$$

where $\hat{\Sigma}_*^{-1}$ and $\hat{\Sigma}_0^{-1}$ are the matrices discussed above with the elements depending on the p_{ij}'s replaced by the appropriate elements of $\tilde{p}_*$, the modal value of the posterior distribution. Last, the approximate posterior variance–covariance matrix for the deleted parameter vector is given by

$$\text{Var}(p_r) = RV(p_*)R' \quad (9.3.7b)$$
$$\doteq R(X_*'\hat{\Sigma}_*^{-1}X_* + \hat{\Sigma}_0^{-1}J)^{-1}R'.$$

9.4. Comparison with some sampling theory results

It is well known[4] that under not very restrictive general conditions, posterior distributions tend to become normal in large samples with

[4] For example, see Jeffreys (1961), Lindley (1965) and Zellner (1969).

mean (and modal value) equal to the maximum likelihood estimate. Further, in large samples, the variance–covariance matrix associated with a posterior distribution can be approximated by the inverse of the estimated information matrix. Heuristically, *when the sample size is large*, the sample information dominates the prior information (as long as this information is not "dogmatic") and the posterior distribution is well approximated by the likelihood function which, under general conditions, will assume a normal form centered at the ML estimates. Therefore, and perhaps not very surprisingly, there is a dovetailing of Bayesian and ML sampling theory results in large samples. However, it should be appreciated that the interpretation of results in the Bayesian and sampling theory approaches is fundamentally different.

It is also interesting to observe that when the parameters of the prior distribution, the a_{ij}'s, are large, the matrix J in (9.3.5) with elements $(a_i - 1)/(a_i + 1)$ will be very nearly an identity matrix. Further, for large a_{ij}'s, the vector f in (9.3.5) with elements $(a_{ij} - r)/(a_i + 1)$ will have elements approximately equal to a_{ij}/a_i, the prior means. Under these conditions, we can obtain the following relation from (9.3.5):

$$\boldsymbol{p}_* = (X_*'\Sigma_*^{-1}X_* + \Sigma_0^{-1})^{-1}(X_*'\Sigma_*^{-1}\boldsymbol{y}_* + \Sigma_0^{-1}\boldsymbol{p}_0), \qquad (9.4.1)$$

where $\boldsymbol{p}_0$ is the prior mean vector for $\boldsymbol{p}_*$ with elements a_{ij}/a_i, $i = 1$, $2, ..., r$ and $j = 1, 2, ..., r - 1$. When sample quantities are employed to obtain estimates of Σ_* and Σ_0^{-1} and these are inserted in (9.4.1), we obtain a conditional modal value for the posterior distribution appropriate for the assumption made about the a_{ij}'s.

The expression in (9.4.1) can also be generated by modifying the Theil–Goldberger (TG) *sampling theory* approach (1960) for introducing prior information in analyses of the standard linear regression model.[5] In the present case, we proceed as follows. The model for the observations is

$$\boldsymbol{y}_* = X_*\boldsymbol{p}_* + \boldsymbol{u}_* \qquad (9.4.2)$$

---

[5] In Tiao and Zellner (1965), it was shown in connection with analyses of the multiple regression model that the Theil–Goldberger approach yields an estimate for the regression coefficient vector which is identical to the mean of a leading normal term in an asymptotic expansion approximating the posterior distribution for the regression coefficient vector.

with

$$Eu_* = 0, \qquad (9.4.3)$$

and

$$Eu_* u'_* = \Sigma_*, \qquad (9.4.4)$$

where Σ_* is assumed non-singular. In the TG approach, prior information is introduced as follows:

$$p_0 = p_* + d \qquad (9.4.5)$$

In (9.4.4), the investigator provides the vector p_0, p_* is regarded as a vector of fixed, unknown parameters, and d is regarded to be a vector of errors with the following properties:

$$Ed = 0, \qquad (9.4.6)$$

and

$$Edd' = \Sigma_0, \qquad (9.4.7)$$

where Σ_0 is a non-singular matrix with elements assigned values by the investigator. Also, it is assumed that

$$Edu' = 0; \qquad (9.4.8)$$

that is, the errors in the equations in (9.4.2) are uncorrelated with the errors in (9.4.5) which characterize the precision of the prior information.

We can now write (9.4.2) and (9.4.5) as

$$\begin{pmatrix} y_* \\ p_0 \end{pmatrix} = \begin{pmatrix} X_* \\ I \end{pmatrix} p_* + \begin{pmatrix} u_* \\ d \end{pmatrix}. \qquad (9.4.9)$$

Written in this way, it is apparent that the TG approach can be viewed as adding additional 'observations', the elements of p_0, to the vector of observations, y_*.[6]

Under the assumptions introduced above, the covariance matrix of the error vector in (9.4.9) is given by

$$\mathrm{Var}\begin{pmatrix} u_* \\ d \end{pmatrix} = \begin{pmatrix} \Sigma_* & 0 \\ 0 & \Sigma_0 \end{pmatrix}. \qquad (9.4.10)$$

---

[6] Note that (9.4.6) implies $Ep_0 = p_*$. That is the additional 'observations' are assumed to have a mean equal to the true parameter p_*. This is a very strong condition on the prior information and one that is not imposed in the Bayesian approach.

Then application of Aitken's generalized least squares procedure to the system in (9.4.9) yields

$$\bar{p}_* = \left[\binom{X_*}{I}' \binom{\Sigma_* \quad 0}{0 \quad \Sigma_0}^{-1} \binom{X_*}{I} \right]^{-1} \left[\binom{X_*}{I}' \binom{\Sigma_* \quad 0}{0 \quad \Sigma_0}^{-1} \binom{y_*}{p_0} \right]$$

$$= (X_*' \Sigma_*^{-1} X_* + \Sigma_0^{-1})^{-1} (X_*' \Sigma_*^{-1} y_* + \Sigma_0^{-1} p_0). \qquad (9.4.11)$$

Note that (9.4.11) is in precisely the same form as (9.4.1).

While $\bar{p}_*$ in (9.4.11) is an optimal estimator in the least squares sense, it must be realized that it depends on the matrix Σ_* whose elements are unknown and depends, as seen above, on the p_{ij}'s, the parameters to be estimated. If, as TG do in the standard regression problem, we substitute a sample estimate for Σ_* in (9.4.11), the resulting estimator has a large sample justification. However, in large samples, the prior information will be dominated by the sample information. Note that elements of $X_*' \Sigma_*^{-1} X_*$ grow in size as the sample size increases whereas the elements of Σ_0^{-1} do not. Finally, it should be noted that the prior information represented by (9.4.5)–(9.4.7) does not include the restrictions $0 \le p_{ij} \le 1$ and $\Sigma_{j=1}^r p_{ij} = 1$, $i = 1, 2, ..., r$. Thus, $\bar{p}_*$, or an approximation to $\bar{p}_*$ based on an estimate of Σ_*, can result in estimates of the transition probabilities which fall in an inadmissible region of the parameter space.

9.5. A macro Bayesian transition probability estimator

The quantity $\overset{\approx}{p}_*$ in (9.3.6a) will be an acceptable approximation to the modal value of the posterior distribution in (9.2.2), provided that it falls in an admissible region of the parameter space; that is, satisfies $0 \le \overset{\approx}{p}_{ij} \le 1$ and the adding up constraints. If this is indeed the case, then we have quantities which can be used as point estimates. On the other hand, it may be that (9.3.6) yields estimates in an inadmissible region of the parameter space. In this event, we now present a procedure for maximizing the posterior probability density function subject to the conditions (9.3.1) and (9.3.2).

Proceeding as before, and thus applying the reducibility theorem of nonlinear programming, we obtain the equivalent problem:

To maximize

$$(X'_* \Sigma_*^{-1} y_* - X'_* \Sigma_*^{-1} X_* \ddot{\bar{p}}_* + \Sigma_0^{-1} f - \Sigma_0^{-1} J \ddot{\bar{p}}_*)' p_*, \qquad (9.5.1)$$

subject to

$$R p_* \leq \eta_r, \qquad (9.5.2)$$

and

$$p_* \geq 0, \qquad (9.5.3)$$

where $R = (I_1 \, I_2 \, \cdots \, I_{r-1})$, each I_i is an $(r \times r)$ identity matrix and η_r is a $(r \times 1)$ column vector with elements unity. By the duality theorem (Dantzig and Orden 1953), if the above primal problem has an optimal solution, so does the following dual problem:
To minimize

$$\lambda' \eta_r \qquad (9.5.4)$$

subject to

$$R' \lambda \geq X'_* \Sigma_*^{-1} y_* + \Sigma_0^{-1} f - X'_* \Sigma_*^{-1} X_* p_* - \Sigma_0^{-1} J p_* \qquad (9.5.5)$$

and

$$\lambda \geq 0, \qquad (9.5.6)$$

where λ is a $(r \times 1)$ column vector of dual variables. By the primal–dual programming formulation, the following problem results:
To maximize

$$(X'_* \Sigma_*^{-1} y_* + \Sigma_0^{-1} f - X'_* \Sigma_*^{-1} X_* p_* - \Sigma_0^{-1} J)' p_* - \lambda' \eta_r$$
$$= -\alpha' p_* - \lambda' p_r \leq 0 \qquad (9.5.7)$$

subject to

$$R p_* + p_r = \eta_r, \qquad (9.5.8)$$

$$R' \lambda + (X'_* \Sigma_*^{-1} X_* + \Sigma_0^{-1} J) p_* - \alpha = X'_* \Sigma_*^{-1} y_* + \Sigma_0^{-1} f, \qquad (9.5.9)$$

and

$$p_*, \lambda, p_r, \alpha \geq 0, \qquad (9.5.10)$$

where p_r and α are $(r \times 1)$ vectors of slack variables. The vector p_r is also a vector of parameters which is deleted in the nonsingular formulation of Σ_*. The simplex tableau for this formulation is given in table 9.1.

When the parameters of the prior matrix beta PDF are large, the prior PDF tends to be multinormal with J approaching I and f approaching p_0. Thus, the simplex tableau becomes table 9.2.

TABLE 9.1

The simplex tableau for the Bayesian estimator with matrix beta prior PDF

B_0	$\lambda \geq 0$	$p_* \geq 0$	$p_r \geq 0$	$\alpha \geq 0$
η_r	0	R	I	0
$X_*'\Sigma_*^{-1}y_* + \Sigma_0^{-1}f$	R'	$X_*'\Sigma_*^{-1}X_* + \Sigma_0^{-1}J$	0	$-I$

TABLE 9.2

The simplex tableau for the Bayesian estimator with multinormal prior PDF

B_0	$\lambda \geq 0$	$p_* \geq 0$	$p_r \geq 0$	$\alpha \geq 0$
η_r	0	R	I	0
$X_*'\Sigma_*^{-1}y_* + \Sigma_0^{-1}p_0$	R'	$X_*'\Sigma_*^{-1}X_* + \Sigma_0^{-1}$	0	$-I$

9.6. The Bayesian approach: further considerations

Above, in (9.2.2), we have presented the posterior distribution of the transition probabilities which incorporates both sample and prior information. For the special case of a (2×2) transition probability matrix, that is the two state case, bivariate numerical integration techniques can readily be employed to obtain the normalized bivariate posterior distribution for p_{11} and p_{21} as well as the marginal posterior PDF's for these parameters. These distributions can be utilized to make probability statements about these two parameters. Further, since $p_{11} + p_{12} = 1$ and $p_{21} + p_{22} = 1$, posterior probability statements can also be made about p_{12} and p_{22}. Then too, if a loss function is given, optimal point estimates can be calculated numerically. For example, if the loss function is quadratic, then as is well known the mean of the posterior distribution is an optimal estimate. For the two state case, the mean of the posterior distribution can be calculated using numerical integration techniques. Thus, for the two state case, a complete analysis of the posterior distribution can be performed.

For systems with more than two states, the number of parameters in the transition matrix is rather large. In such situations, current numerical integration techniques known to the authors do not appear capable of providing a convenient approach for analyzing the posterior distribution. In view of this situation and the fact that the posterior distribution is rather complicated, we have just presented an expression for the modal value of the posterior distribution and an approximate expression for the variance–covariance matrix. Further analytical and numerical procedures for analyzing the posterior distribution are required.

The analysis presented in earlier sections of this chapter reveals that the modal values of the posterior distribution for the transition probabilities can be viewed as a matrix weighted average of sample and prior information – see in particular, (9.3.6a). The matrices Σ_*^{-1} and Σ_0^{-1}, the inverses of the disturbance and prior covariance matrices respectively, play an important role in weighting the sample and prior information concerning the location of the transition probabilities. As usual in Bayesian analysis, if our prior PDF has small dispersion, the posterior distribution's location will be importantly determined in small to moderate sized samples by the values assigned to the prior means of the transition probabilities. On the other hand, if the prior PDF is quite spread out, then the sample information will play a greater role in determining the location of the posterior distribution. If, for example, a uniform prior distribution were used for the transition probabilities, the posterior distribution would be proportional to the likelihood function and, of course, the modal value of the posterior distribution would coincide with the maximum likelihood estimate.[7] In large samples, sample information dominates the posterior distribution and thus the form of the prior, as long as it is not "dogmatic", has little influence in determining the properties of the posterior distribution.

In many instances investigators wish to determine how the information in a sample of data modifies prior beliefs about transition probabilities. If an investigator can represent his prior information by use of the multivariate beta PDF in (9.1.1), he can combine this prior PDF with the likelihood function and compare properties of the resulting posterior distribution with those of his prior distribution. In addition, if an investigator decides to introduce his prior information in his analysis of a sample of data, then posterior inferences will incorporate both the sample and prior information as explained above.

---

[7] The mean of the posterior distribution, an optimal point estimate for a quadratic loss function, may depart markedly from the maximum likelihood estimate in small samples.

9.7. A numerical example

To show the numerical computation in detail, the simple absorbing chain used in ch. 7 will be used. The aggregate data (7.5.1) are used to obtain the cross products $X'_* \Sigma_*^{-1} X_*$ and $X'_* \Sigma_*^{-1} y_*$, which were given by (7.5.3) and (7.5.4) in ch. 7.

The multivariate beta PDF for this 2-Markov-state case becomes a univariate beta PDF. The prior parameters are listed in table 9.3.

TABLE 9.3
The list of the prior parameters for beta distributions

p_{ij}	a_{ij}	$a_i - a_{ij}$	Mean	Variance
p_{11}	99	1	0.9900	0.0001
p_{12}	1	99	0.0100	0.0001
p_{21}	50	50	0.5000	0.0025
p_{22}	50	50	0.5000	0.0025

Thus, the cross products for sample and prior information are

$$X'_* \Sigma_*^{-1} X_* + \Sigma_0^{-1} = \begin{bmatrix} 5075.4790 & 691.9386 \\ 691.9386 & 230.4220 \end{bmatrix} + \begin{bmatrix} 9898.9900 & 0 \\ 0 & 392.0000 \end{bmatrix}$$

$$= \begin{bmatrix} 14974.4690 & 691.9386 \\ 691.9386 & 622.4220 \end{bmatrix} \tag{9.7.1}$$

and

$$X'_* \Sigma_*^{-1} X_* + \Sigma_0^{-1} p_0 = \begin{bmatrix} 5425.00 \\ 808.33 \end{bmatrix} + \begin{bmatrix} 9898.99 & 0 \\ 0 & 392.00 \end{bmatrix} \begin{bmatrix} 0.99 \\ 0.50 \end{bmatrix}$$

$$= \begin{bmatrix} 15323.99 \\ 1004.33 \end{bmatrix}. \tag{9.7.2}$$

The unrestricted estimates are then

$$\begin{bmatrix} \dot{p}_{11} \\ \dot{p}_{21} \end{bmatrix} = \begin{bmatrix} 0.00007 & -0.00008 \\ -0.00008 & 0.00169 \end{bmatrix} \begin{bmatrix} 15323.99 \\ 1004.33 \end{bmatrix} = \begin{bmatrix} 1.0002 \\ 0.5017 \end{bmatrix}, \tag{9.7.3}$$

and

$$\begin{bmatrix} \dot{p}_{12} \\ \dot{p}_{22} \end{bmatrix} = \begin{bmatrix} 1 \\ 1 \end{bmatrix} - \begin{bmatrix} 1.0002 \\ 0.5017 \end{bmatrix} = \begin{bmatrix} -0.0002 \\ 0.4933 \end{bmatrix}. \tag{9.7.4}$$

The unrestricted estimates violates the non-negativity restriction and hence, the iteration procedure is required. In the quadratic iterations, p_{12} and p_{22} are introduced into the basis initially. After exactly three iterations, p_{11} replaces p_{12} and p_{21} and α_1 takes over for the artificial variables and the solution is optimal. The resulting Bayesian estimates are then

$$\hat{\tilde{P}}^c = \begin{bmatrix} 1 & 0 \\ 0.5019 & 0.4981 \end{bmatrix}. \qquad (9.7.5)$$

9.8. Sampling experiment results

The amended aggregate data from period 3 to 14 are used for the sampling experiment. The sample information $X_*'\Sigma_*^{-1}X_*$ and $X_*'\Sigma_*^{-1}y_*$ is, in fact, exactly the same as that of the ML estimator (also is equivalent to that of the GLS estimator presented in ch. 6). The prior knowledge about the multivariate beta distribution is given in table 9.4.

TABLE 9.4

The list of the parameters for the multi-beta prior distribution assuming that the information is given by the behavior of the micro units

p_{ij}	a_{ij}	$a_i - a_{ij}$	Mean	Variance
p_{11}	599	401	0.5990	0.00023995
p_{12}	399	601	0.3990	0.00023956
p_{13}	1	999	0.0010	0.00000099
p_{14}	1	999	0.0010	0.00000099
p_{21}	100	900	0.1000	0.00008991
p_{22}	499	501	0.4990	0.00024974
p_{23}	400	600	0.4000	0.00023976
p_{24}	1	999	0.0010	0.00000099
p_{31}	1	999	0.0010	0.00000099
p_{32}	100	900	0.1000	0.00008991
p_{33}	699	301	0.6990	0.00021019
p_{34}	200	800	0.2000	0.00014984
p_{41}	1	999	0.0010	0.00000099
p_{42}	1	999	0.0010	0.00000099
p_{43}	100	900	0.1000	0.00008991
p_{44}	898	102	0.8980	0.00009150

In table 9.4, we have assigned $a_i = 1000$ because we assume that we know the behavior of the 1000 individuals we have generated from the simulation model presented in ch. 4. The variances of the prior distributions are small and we expect a larger influence from the prior than the sample information.

a. Multi-beta leptokurtic prior

As an initial experiment, the covariances of the prior distributions are ignored and we assume that each p_{ij} has an independent beta prior PDF. The means and root mean square errors for 50 unrestricted Bayesian estimates from samples of size 25, 50 and 100 are given in table 9.5.

TABLE 9.5

The means and root mean square errors of the unrestricted Bayesian estimates with leptokurtic multi-beta prior (sample periods from 3 to 14)

Sample size	Means				Root mean square error			
25	0.6047	0.3986	−0.0000	0.0027	0.0428	0.0024	0.0000	0.0045
	0.0990	0.4983	0.4002	0.0025	0.0017	0.0040	0.0030	0.0069
	0.0000	0.0982	0.7001	0.2018	0.0000	0.0026	0.0032	0.0044
	0.0000	0.0000	0.0991	0.9009	0.0000	0.0000	0.0015	0.0015
50	0.5995	0.4004	0.0000	0.0000	0.0035	0.0032	0.0000	0.0060
	0.0987	0.5000	0.4024	−0.0011	0.0024	0.0047	0.0050	0.0087
	0.0000	0.0979	0.7010	0.2012	0.0000	0.0043	0.0040	0.0054
	0.0000	0.0000	0.0994	0.9006	0.0000	0.0000	0.0013	0.0013
100	0.6007	0.4008	0.0000	−0.0015	0.0039	0.0035	0.0000	0.0061
	0.0991	0.5007	0.4035	−0.0033	0.0030	0.0047	0.0056	0.0092
	0.0000	0.0989	0.7012	0.1999	0.0000	0.0026	0.0036	0.0044
	0.0000	0.0000	0.0991	0.9009	0.0000	0.0000	0.0017	0.0018

The unrestricted Bayesian estimates are better than the ML estimates or GLS estimates in the sense there are only a few negative estimates. Although the means of the samples of size 25 are all positive, this does not mean that all the estimates are acceptable. The aggregate errors measured by the sum of all the root mean square error elements are

0.0383, 0.0500 and 0.0503 for samples of size 25, 50 and 100 respectively. They are relatively small when compared to the ML or GLS estimates which have an aggregate root mean square error of 3.6036, 2.9800 and 2.7936 respectively. It is interesting that when sample size increases, the estimates do not improve very much. This is, of course, due to the supremacy of the leptokurtic prior distribution. The probability is almost determined by the prior instead of the likelihood and hence is the extreme opposite case of stable estimation. Therefore, it is dangerous to assign a small variance prior unless the prior knowledge is securely based.

Since the unrestricted estimates are not necessarily acceptable, we have also computed the restricted Bayesian estimates for three levels of size 25, 50 and 100. The means and root mean square errors are given in table 9.6.

TABLE 9.6

The means and root mean square errors of the restricted Bayesian estimates with leptokurtic multi-beta prior (sample periods from 3 to 14)

Sample size	Means				Root mean square error			
25	0.5984	0.3983	0.0000	0.0032	0.0025	0.0024	0.0000	0.0043
	0.0987	0.4997	0.3995	0.0041	0.0016	0.0037	0.0024	0.0058
	0.0000	0.0982	0.7002	0.2016	0.0000	0.0026	0.0034	0.0046
	0.0000	0.0000	0.0989	0.9009	0.0000	0.0000	0.0019	0.0015
50	0.5984	0.3993	0.0000	0.0023	0.0028	0.0020	0.0000	0.0039
	0.0981	0.4982	0.4006	0.0031	0.0027	0.0039	0.0028	0.0054
	0.0000	0.0981	0.7016	0.2003	0.0000	0.0043	0.0048	0.0063
	0.0000	0.0000	0.0994	0.9006	0.0000	0.0000	0.0012	0.0013
100	0.5991	0.3992	0.0000	0.0016	0.0029	0.0022	0.0000	0.0031
	0.0984	0.4985	0.4012	0.0019	0.0032	0.0037	0.0027	0.0036
	0.0000	0.0995	0.7025	0.1980	0.0000	0.0027	0.0051	0.0065
	0.0000	0.0000	0.0092	0.9008	0.0000	0.0000	0.0017	0.0017

On all tests applied, the restricted estimates are better than the unrestricted estimates. The aggregate root mean square errors are 0.0365, 0.0415 and 0.0391 respectively for samples of size 25, 50 and 100.

The tests of normality by the Kolmogorov–Smirnov D statistic show that none of the elements of the unrestricted estimates has a significant deviation from the normal distribution at the 10 percent significance level. The restricted estimates also have no significant deviation from the normal distribution except those elements whose target values are zero. The percentages hitting the target value for these elements are given in table 9.7. The percentages of hitting the target value are more evenly distributed for the sample size of 100 than for 50 and 25.

TABLE 9.7

The percentage of hitting the target value for the restricted multi-beta prior Bayesian estimator

Elements	Sample size		
	25	50	100
	(%)	(%)	(%)
p_{13}	100	98	100
p_{14}	64	52	64
p_{24}	36	52	64
p_{31}	100	100	94
p_{41}	100	100	100
p_{42}	100	100	100

b. Multivariate beta leptokurtic prior

In the second experiment, covariances of the prior distributions are not ignored and a *multivariate* beta prior is used. The means and root mean square errors for these 50 unrestricted Bayesian estimates from samples of sizes 25, 50 and 100 are given in table 9.8.

The aggregate errors as measured by the sum of all the root mean square error elements are 0.0233, 0.0326 and 0.0335 for samples of size 25, 50 and 100 respectively. These results are slightly superior to those obtained in the multi-beta experiment but in most respects, given the supremacy of the prior, the results are identical.

Since a few of the unrestricted estimates violated the non-negativity condition, the restricted multivariate beta prior estimates are given in table 9.9.

TABLE 9.8

The mean and root mean square errors of the unrestricted Bayesian estimates with leptokurtic multivariate-beta prior

Sample size	Means				Root mean square error			
25	0.6001	0.3999	0.6000	0.0000	0.0021	0.0021	0.0000	0.0000
	0.0993	0.4992	0.4015	0.0000	0.0013	0.0035	0.0034	0.0000
	0.0000	0.0981	0.7022	0.1997	0.0000	0.0026	0.0038	0.0022
	0.0000	−0.0000	0.0993	0.9007	0.0000	0.0000	0.0011	0.0011
50	0.5995	0.4005	0.0000	−0.0000	0.0026	0.0026	0.0000	0.0000
	0.0986	0.4991	0.7033	0.1990	0.0024	0.0039	0.0042	0.0000
	−0.0000	0.0976	0.7033	0.1990	0.0000	0.0045	0.0060	0.0035
	−0.0000	−0.0000	0.0995	0.9005	0.0000	0.0001	0.0011	0.0012
100	0.6000	0.3999	0.0000	−0.0000	0.0037	0.0037	0.0000	0.0000
	0.0988	0.4989	0.4023	−0.0000	0.0030	0.0043	0.0039	0.0000
	−0.0000	0.0987	0.7029	0.1983	0.0000	0.0024	0.0048	0.0039
	−0.0000	−0.0000	0.0993	0.9007	0.0000	0.0000	0.0015	0.0016

TABLE 9.9

The means and root mean square errors of the restricted Bayesian estimates with a leptokurtic multivariate-beta prior

Sample size	Means				Root mean square error			
25	0.6001	0.3999	0.0000	0.0000	0.0021	0.0021	0.0000	0.0000
	0.0993	0.4992	0.4015	0.0000	0.0013	0.0035	0.0034	0.0000
	0.0000	0.0981	0.7022	0.1997	0.0000	0.0026	0.0038	0.0022
	0.0000	0.0000	0.0993	0.9006	0.0000	0.0000	0.0011	0.0011
50	0.5995	0.4005	0.0000	0.0000	0.0026	0.0026	0.0000	0.0000
	0.0986	0.4990	0.4023	0.0000	0.0024	0.0039	0.0042	0.0000
	0.0000	0.0977	0.7033	0.1990	0.0000	0.0045	0.0061	0.0036
	0.0000	0.0000	0.0995	0.9005	0.0000	0.0000	0.0011	0.0012
100	0.6000	0.3999	0.0000	0.0000	0.0037	0.0037	0.0000	0.0000
	0.0988	0.4989	0.4023	0.0000	0.0030	0.0043	0.0039	0.0000
	0.0000	0.0987	0.7029	0.1933	0.0000	0.0025	0.0048	0.0040
	0.0000	0.0000	0.0993	0.9007	0.0000	0.0000	0.0016	0.0016

The aggregate root mean square errors for this experiment are 0.0232, 0.0325 and 0.0334 for samples of size 25, 50 and 100 respectively. The results differ only very slightly from the unrestricted estimates and are, of course, very similar to the multi-beta experiment results. All of the test results for the multi-beta experiments also apply to those reported in tables 9.8 and 9.9.

c. Results from a platykurtic prior

In the case when very little prior knowledge is given, the prior distribution should have a large variance. To reflect this situation, we performed an experiment assuming very little prior information. Suppose that we know only that the parameters will always lie between zero and one. We may assign the parameters of the multi-beta distribution such that the means for all transition probabilities are equal; that is, the micro units have an equal chance to transit to any other Markov state. We may also assign the variance of the prior distribution to be very large such that three standard deviations will cover almost the entire

TABLE 9.10

Tests of the effect of alternative prior parameters with a fixed mean on the Bayesian estimator (sample information from sample 6 of size 50)

	Beta prior				Chi-square		MSE	
a_{ij}	$a_i - a_{ij}$	Mean	Variance	σ	Unrest.	Rest.	Unrest.	Rest.
1.5	4.5	0.25	0.0269	0.16	13.5565	16.6946	0.0012	0.0014
1.4	4.2	0.25	0.0284	0.17	13.5084	16.5379	0.0012	0.0014
1.3	3.9	0.25	0.0302	0.17	13.4034	16.3857	0.0011	0.0014
1.2	3.6	0.25	0.0323	0.18	13.4223	16.2413	0.0011	0.0014
1.1	3.3	0.25	0.0347	0.18	13.3863	16.1094	0.0011	0.0013
1.0	3.0	0.25	0.0375	0.19	13.3575	15.9963	0.0011	0.0013
0.9	2.7	0.25	0.0408	0.20	13.3395	15.9105	0.0011	0.0013
0.8	2.4	0.25	0.0446	0.21	13.3388	15.8638	0.0011	0.0013
0.7	2.1	0.25	0.0493	0.22	13.3692	15.8721	0.0011	0.0013
0.6	1.8	0.25	0.0551	0.23	13.4617	15.9664	0.0010	0.0013
0.5	1.5	0.25	0.0625	0.25	13.7078	15.2290	0.0010	0.0013
0.4	1.2	0.25	0.0721	0.27	23.0324	16.6977	0.0018	0.0013
0.3	0.9	0.25	0.0952	0.29	17.2145	19.0623	0.0013	0.0014
0.2	0.6	0.25	0.1042	0.32	506.3499	19.8018	0.0032	0.0014
0.1	0.3	0.25	0.1339	0.37	153.3647	21.9741	0.1087	0.0017

range of the distribution. Before proceeding to the large scale sampling experiment, sample 6 of size 50 was chosen randomly. Several cases of different parameters are tested and the results are shown in table 9.10.

In table 9.10, the chi-square test reported is defined in ch. 7 and MSE denotes the mean square error of the predicted proportions. For sample 6 of size 50, when the parameters $a_{ij} = 0.8$ and $a_i - a_{ij} = 2.4$, the chi-square value is the smallest. The large scale experiment was performed with the prior $a_{ij} = 0.8$, $a_i - a_{ij} = 2.4$. The experimental results are given in table 9.11.

TABLE 9.11

The means and root mean square errors of the unrestricted Bayesian estimates with platykurtic multi-beta prior (sample periods from 3 to 14)

Sample size	Means				Root mean square error			
25	0.3467	0.4291	0.1864	0.0376	0.2872	0.1477	0.2657	0.2055
	0.1701	0.4280	0.4205	−0.0186	0.1158	0.1530	0.1398	0.1363
	0.0473	0.1461	0.5474	0.2592	0.0927	0.1049	0.2028	0.1272
	−0.0189	0.0512	0.1939	0.8761	0.0736	0.0811	0.1447	0.1013
50	0.4087	0.4453	0.1407	0.0053	0.2194	0.1421	0.2111	0.1419
	0.1788	0.4259	0.4345	−0.0392	0.1266	0.1463	0.1363	0.1487
	0.0214	0.1625	0.5453	0.2709	0.0904	0.1344	0.1935	0.1310
	−0.0208	−0.0556	0.2232	0.8532	0.0769	0.1069	0.1562	0.1061
100	0.4342	0.4447	0.1216	−0.0005	0.1906	0.1286	0.1980	0.1541
	0.2166	0.3824	0.4208	−0.0198	0.1455	0.1510	0.1242	0.1652
	−0.0080	0.1935	0.5930	0.2215	0.0651	0.1259	0.1319	0.0832
	−0.0120	0.0561	0.1752	0.8928	0.0490	0.0850	0.1049	0.0627

Table 9.11 shows that the platykurtic prior gives little weight to the prior knowledge and that the estimated matrices still have the form of the specified transition probability model in spite of subjective prior means of 0.25 for each transition probability. The property of consistency still persists in that the aggregate root mean square errors are 2.3793, 2.2678 and 1.9659 for sample size 25, 50 and 100 respectively.

The results of the restricted estimates are given in table 9.12. The asymptotic tendency of the estimates is to be closer to the true para-

TABLE 9.12

The means and root mean square errors of the restricted Bayesian estimates
with platykurtic multi-beta prior (sample periods from 3 to 14)

Sample size	Means				Root mean square error			
25	0.3631	0.4261	0.1841	0.0268	0.2820	0.1169	0.2176	0.0524
	0.1578	0.4403	0.3952	0.0067	0.0967	0.1366	0.1097	0.0178
	0.0305	0.0995	0.6244	0.2456	0.0464	0.0709	0.1417	0.1066
	0.0089	0.0021	0.1288	0.8602	0.0208	0.0079	0.0960	0.0972
50	0.4403	0.4436	0.1126	0.0036	0.1865	0.1277	0.1448	0.0108
	0.1418	0.4607	0.3918	0.0057	0.0958	0.1223	0.0882	0.0166
	0.0115	0.0976	0.6696	0.2213	0.0259	0.0575	0.0923	0.0711
	0.0035	0.0035	0.1185	0.8745	0.0081	0.0129	0.0683	0.0715
100	0.4808	0.4266	0.0913	0.0913	0.1494	0.1116	0.1226	0.0050
	0.1518	0.4531	0.3911	0.0041	0.0859	0.0999	0.0761	0.0113
	0.0047	0.1107	0.6882	0.1963	0.0119	0.0392	0.0526	0.0393
	0.0004	0.0023	0.0986	0.8988	0.0015	0.0082	0.0513	0.0510

meters as the sample size increases and the aggregate errors are 1.5973, 1.2003 and 0.9167 for sample sizes 25, 50 and 100 respectively.

From the above experiments, we may conclude that if the prior knowledge is available, we should take advantage of the information and apply Bayes' theorem to obtain the posterior distribution for use in decision making. If the prior knowledge is very complete, we may assign the subjective prior with a small variance to provide a large weight on the prior. If the prior knowledge is not very reliable, we may assign the subjective prior with a large variance to provide a small weight on the prior reflecting the principle of stable estimation.

9.9. Results for the brand change problem

The previous results for the Telser problem have only used sample information. Assume now that the following prior knowledge is available:

(1) There is an equal chance for a cigarette smoker to shift to another brand or to remain consuming a particular brand of cigarette.
(2) If a smoker shifts, he will also have the same chance of choosing any other brand of cigarettes. Under this specification, the prior means about the transition probabilities are:

$$
\begin{array}{c}
\begin{array}{ccc} \text{Camels} & \text{Lucky} & \text{Chester-} \\ & \text{Strike} & \text{field} \end{array} \\
\begin{array}{l} \text{Camels} \\ \text{Lucky Strike} \\ \text{Chesterfield} \end{array}
\left[\begin{array}{ccc} 1/2 & 1/4 & 1/4 \\ 1/4 & 1/2 & 1/4 \\ 1/4 & 1/4 & 1/2 \end{array} \right].
\end{array}
\tag{9.9.1}
$$

If the degree of belief about the transition matrix is not very certain, the multivariate-beta prior covariance matrix should be assigned with large values relative to their means. Assume that the covariance matrix is given by table 9.13.

TABLE 9.13
Prior covariance matrix of transition probabilities for cigarette consumers

	p_{11}	p_{21}	p_{31}	p_{12}	p_{22}	p_{32}
p_{11}	0.0500			−0.0250		
p_{21}		0.0375			−0.0250	
p_{31}			0.0375			−0.0250
p_{12}	−0.0250			0.0500		
p_{22}		−0.0250			0.0375	
p_{32}			−0.0250			0.0375

From the means and the covariances, the prior parameters a_{ij} of the multivariate-beta probability density function are:

$$
[a_{ij}] = \left[\begin{array}{ccc} 2 & 1 & 1 \\ 1 & 2 & 1 \\ 1 & 1 & 2 \end{array} \right].
\tag{9.9.2}
$$

The incorporation of the above prior knowledge with the sample observation yields the Bayesian estimates shown in table 9.14.

TABLE 9.14

Bayesian estimates of the transition matrix of the cigarette consumers by recursive quadratic programming for three leading brands: Camels, Lucky Strike and Chesterfield

Stages	Estimates			Recursive difference
Initial solution	$\begin{bmatrix} 0.7802 \\ 0 \\ 0.2545 \end{bmatrix}$	$\begin{matrix} 0.1251 \\ 0.8910 \\ 0 \end{matrix}$	$\begin{matrix} 0.0947 \\ 0.1090 \\ 0.7455 \end{matrix}$	—
1st recursive solution	$\begin{bmatrix} 0.7830 \\ 0 \\ 0.2517 \end{bmatrix}$	$\begin{matrix} 0.1220 \\ 0.8944 \\ 0 \end{matrix}$	$\begin{matrix} 0.0950 \\ 0.1056 \\ 0.7485 \end{matrix}$	0.0112
2nd recursive solution	$\begin{bmatrix} 0.7835 \\ 0 \\ 0.2509 \end{bmatrix}$	$\begin{matrix} 0.1216 \\ 0.8948 \\ 0 \end{matrix}$	$\begin{matrix} 0.0949 \\ 0.1052 \\ 0.7491 \end{matrix}$	0.0022
3rd recursive solution	$\begin{bmatrix} 0.7836 \\ 0 \\ 0.2508 \end{bmatrix}$	$\begin{matrix} 0.1216 \\ 0.8948 \\ 0 \end{matrix}$	$\begin{matrix} 0.0948 \\ 0.1052 \\ 0.7492 \end{matrix}$	0.0004

CHAPTER 10

THE MINIMUM ABSOLUTE DEVIATIONS ESTIMATOR

10.1. Specification of the statistical model

To estimate the transition probabilities for a Markov chain, we fit a set of regression hyperplanes, which we may write in the multivariate form

$$y = Xp + u, \tag{10.1.1}$$

for which no intercepts are involved. An alternative criterion for deriving parameter estimates when equality or inequality restrictions are present is to make use of a sum of absolute deviations, perhaps weighted, rather than a sum of squared deviations. Use of the former criterion yields a minimum absolute deviations (MAD) estimator (Fisher 1961; Wagner 1959). If we use the method of minimizing the sum of the absolute deviations, our problem may be specified as follows: find the vector $\bar{p}$ such that the sum of the absolute deviations

$$a'\eta_{rT} \tag{10.1.2}$$

is minimized subject to the constraints

$$y = Xp + u, \tag{10.1.3}$$

$$Gp = \eta_r, \tag{10.1.4}$$

and

$$p \geq 0, \tag{10.1.5}$$

where $a' = (u_1(t), u_2(t), ..., u_r(t))$, a $(1 \times rT)$ vector with $u_i(t) = [|u_i(1)|, |u_i(2)|, ..., |u_i(T)|]$, u is an error term allowing either positive or negative values and η_{rT} and η_r are unit vectors with dimension rT and r respectively. Given the observations y and X, the model (10.1.2) to (10.1.5) may be solved by linear programming.

131

10.2. Linear programming formulation

To formulate our estimation problem a linear programming problem, we will have to modify the deviation vector u so that it appears in both the objective function (10.1.2) and the constraints (10.1.3) in the same manner. We will denote the absolute value $|u_j(t)|$ by the following expression:

$$|u_j(t)| = \max (0, u_j(t)) + \max (-u_j(t), 0). \qquad (10.2.1)$$

Then, the value of $u_j(t)$ itself may be written as

$$u_j(t) = \max (0, u_j(t)) - \max (-u_j(t), 0), \qquad (10.2.2)$$

which may be positive or negative. Let u_1 be the vector of max $(0, u_j(t))$ and u_2 be the vector of max $(-u_j(t), 0)$. The resulting u_1 and u_2 have the properties that they are non-negative and that their corresponding elements are counterparts, in the sense that both elements cannot take positive values at the same time. In other words, allowing either positive or negative values for u, we form a pair of counterparts u_1 and u_2 such that u_1 and u_2 are all non-negative and u_1 stands for positive deviations and u_2 for negative deviations.

a. Minimizing the unweighted sum of the absolute deviations

If the minimization criterion is that the unweighted sum of the absolute deviations is to be minimzed, then the linear programming problem is: To minimize

$$(u_1 + u_2)' \, \eta_{rT}, \qquad (10.2.3)$$

subject to

$$y = Xp + u_1 - u_2, \qquad (10.2.4)$$

$$Gp = \eta_r, \qquad (10.2.5)$$

and

$$p, u_1, u_2 \geq 0. \qquad (10.2.6)$$

The problem may be solved by employing the simplex algorithm, and the simplex tableau in the form minimization problem is given in table 10.1.

In the tableau, the error term u_1 may be introduced initially into the basis as a short cut to reduce the rounding error due to the iterations.

TABLE 10.1

The linear programming simplex tableau for the minimum absolute deviations estimator

C_j		0	η_{rT}	η_{rT}
	B_0	p	u_1	u_2
η_{rT} y		X	I	$-I$
m η_r		G	0	0

The last constraints $Gp = \eta_r$ must have artificial variables as bases. The m stands for the high cost of an artificial variable being in the basis.

b. Minimizing the weighted sum of the absolute deviations

The objective function of the previous formulation is an unweighted sum of the absolute deviations. A different weight to each observation may be desired if the sampling error in each observation is not uniform. Just as in the weighted least squares formulation, when we used $H'H$ as the weight matrix in minimizing the squared loss function,

$$\phi(p) = (y - Xp) \, H'H(y - Xp)', \qquad (10.2.7)$$

we may wish to use H as a weight matrix in the absolute sum loss function

$$\psi(p) = (Ha)' \, \eta_{rT}. \qquad (10.2.8)$$

One possible weight matrix H may be defined as

$$H'H = \Sigma^{-1}, \qquad (10.2.9)$$

where Σ is the covariance matrix of u.

If u is heteroscedastic but not autocorrelated, then H is the diagonal matrix of the reciprocal of the standard deviations. Thus, to provide smaller weight to observations with large standard deviations and greater weight to observations with small deviations, we divide observations by their standard deviations and then apply the procedure of the previous section. That is, we multiply H through the hyperplanes to obtain

$$Hy = HXp + Hu, \qquad (10.2.10)$$

and then minimize

$$Ha'\eta_{rT} \tag{10.2.11}$$

or

$$(u_1 + u_2)' H'\eta_{rT}, \tag{10.2.12}$$

subject to (10.2.4) through (10.2.6). This implies that the so-called 'cost vector' of the problem becomes $H'\eta_{rT}$ instead of η_{rT}. The simplex tableau for the problem is given in table 10.2.

TABLE 10.2

The linear programming simplex tableau for the weighted MAD *estimator*

C_j		0	$H'\eta_{rT}$	$H'\eta_{rT}$
	B_0	p	u_1	u_2
$H'\eta_{rT}$ y		X	I	$-I$
m η_r		G	0	0

It should be understood that the vectors $H'\eta_{rT}$, p, u_1 and u_2, when appearing in the B_0 and C_j rows of the simplex tableau, are in the transposed form.

10.3. Results from the sampling experiment

The original aggregate data from periods 1 to 16 are used for the experiments. For each problem, the size of the simplex tableau is 64 by 136; that is, there are 136 variables (excluding artificial variables) and 64 constraints. Fifty problems for each sample size of 25, 50 and 100 are computed. For each problem, it requires on the average about 90 iterations and takes about 60 seconds of computing time with the IBM 7094 computer.

The means and root mean square errors statistics for the minimum absolute deviations estimator for each set of 50 samples are presented in table 10.3.

As sample size increases, there is significant improvement in both the means of the point estimates and the root mean square errors with elements of the root mean square error matrices summing to 1.5044,

1.1978 and 0.0453 respectively. The experimental sampling results thus appear to reflect the property of consistency. However, the MAD estimator appears inferior to any of the alternative estimators except unrestricted least squares estimator, when the aggregate mean square er-

TABLE 10.3

Means and root mean square errors for the minimum absolute deviations estimator

Sample size	Means				Root mean square error			
25	0.5096	0.4059	0.0784	0.1481	0.1481	0.1661	0.1500	0.0154
	0.1155	0.4962	0.3701	0.0126	0.0957	0.1554	0.1578	0.0324
	0.0222	0.0864	0.6357	0.2357	0.0408	0.0642	0.1147	0.0948
	0.0044	0.0244	0.1195	0.8517	0.0136	0.0431	0.1057	0.1066
50	0.5633	0.3905	0.0558	0.0004	0.1028	0.1416	0.1078	0.0030
	0.1059	0.5110	0.3788	0.0043	0.0743	0.1455	0.1254	0.0125
	0.0094	0.0840	0.6652	0.2354	0.0220	0.0578	0.1049	0.0811
	0.0052	0.0191	0.1186	0.8571	0.0108	0.0388	0.0824	0.0870
100	0.5810	0.3770	0.0426	0.0000	0.0847	0.1227	0.0910	0.0000
	0.1063	0.5050	0.3876	0.0012	0.0716	0.1343	0.1103	0.0062
	0.0035	0.0947	0.6906	0.2111	0.0094	0.0515	0.0726	0.0487
	0.0037	0.0110	0.0971	0.8882	0.0081	0.0242	0.0552	0.0547

rors are compared. This outcome agrees with the results of a simulation study for the MAD estimator by Ashar and Wallace (1963). However, the method does take into account the restrictions on the transition probabilities.

The weighted MAD estimators have not been computed since the MAD estimator is believed to be less efficient than the estimators previously discussed and the time for computing MAD estimates is considerable.[1]

The hypothesis that the MAD estimates are normally distributed was tested using the Kolmogorov–Smirnov D statistic. This statistic was

[1] With the IBM 7094, it takes about 60 seconds to compute each set of MAD estimates for each MAD estimator, while a restricted ML estimator plus a restricted Bayesian estimator need only 12 seconds.

computed for the estimates of the non-zero elements of the transition matrix for sample sizes of 25, 50 and 100. The computed values of the D statistic are listed in table 10.4.

TABLE 10.4

Kolmogorov–Smirnov D statistic for MAD *estimates*

Sample size	D statistic			
25	0.0992	0.0818	—	—
	0.1725	0.0636	0.0832	—
	—	0.0843	0.0807	0.1354
	—	—	0.1575	0.1022
50	0.0887	0.0776	—	—
	0.1103	0.0959	0.0723	—
	—	0.0738	0.0691	0.1208
	—	—	0.0823	0.1019
100	0.1036	0.0856	—	—
	0.1077	0.0682	0.1452	—
	—	0.0624	0.0910	0.0707
	—	—	0.0603	0.0761

The values of the D statistic in table 10.4 are consistent with the hypothesis that the MAD estimates are normally distributed (5 percent level of significance critical point is 0.18) with the exception of estimate p_{21} for sample size 25. In this case, $D = 0.1725$, a value which is close to the 10 percent critical value ($D = 0.17$). For the estimates such as $p_{13}, p_{14}, p_{24}, p_{41}$ and p_{42}, the percentage hitting the target value (zero) are given in table 10.5. According to table 10.5, the percentage of estimates hitting the target value increases when sample sizes increase, reflecting the property of consistency. The estimates, such as p_{14} of size 100, have one hundred percent chance of hitting the target value. Sample standard deviations computed from the estimates generated in the experiment are given in table 10.6. These standard deviations decrease as the sample size increases. Further, the standard deviations reported in table 10.6 are somewhat larger than those of the other restricted estimators.

TABLE 10.5
*The percentage of hitting the target value
for some elements of the* MAD *estimator*

Elements	25 (%)	50 (%)	100 (%)
p_{13}	52	56	58
p_{14}	98	98	100
p_{24}	78	84	94
p_{31}	58	68	78
p_{41}	82	64	68
p_{42}	50	58	70

TABLE 10.6
Standard deviations of the MAD *estimators*

Sample size	Standard deviations
25	$\begin{bmatrix} 0.1173 & 0.1160 & 0.1279 & 0.0152 \\ 0.0944 & 0.1554 & 0.1549 & 0.0299 \\ 0.0342 & 0.0627 & 0.1058 & 0.0879 \\ 0.0129 & 0.0365 & 0.1039 & 0.0951 \end{bmatrix}$
50	$\begin{bmatrix} 0.0961 & 0.1403 & 0.0923 & 0.0030 \\ 0.0740 & 0.1450 & 0.1236 & 0.0118 \\ 0.0199 & 0.0555 & 0.0990 & 0.0729 \\ 0.0094 & 0.0337 & 0.0802 & 0.0757 \end{bmatrix}$
100	$\begin{bmatrix} 0.0825 & 0.1206 & 0.0804 & 0.0000 \\ 0.0713 & 0.1342 & 0.1096 & 0.0061 \\ 0.0087 & 0.0512 & 0.0720 & 0.0474 \\ 0.0072 & 0.0216 & 0.0552 & 0.0534 \end{bmatrix}$

In conclusion, this study suggests that when the usual criteria for
gauging the performance of an estimator in repeated trials are employed,
the MAD estimator is inferior to the other estimator previously considered. However, the method does take into account the restrictions
on the probabilities, and from the results it appears that the MAD estimator does appear to provide a 'satisfactory' basis for estimating the
transition probabilities.

CHAPTER 11

PREDICTION AND THE CHI-SQUARE
GOODNESS-OF-FIT TEST

One of the objectives of estimating the parameters of an economic
model is to provide a basis for predicting the outcomes of future economic events. In this chapter, we will discuss the prediction of the proportions of the future attributes and a test of the goodness-of-fit of the
model. The goodness-of-fit of the model for the sample data provides
one basis for assessing performance on a wider set of data.

11.1. Predicted proportions

The Markov relation as defined in ch. 1 is given by

$$q_j(t) = \sum_{i=1}^{r} y_i (t - 1) p_{ij}, \qquad (11.1.1)$$

where $y_i (t - 1)$ is the observed proportion and $q_j(t)$ is the true proportion. By introducing an error or disturbance term defined by

$$y_j(t) = q_j(t) + u_j(t), \qquad (11.1.2)$$

the following model for the observed proportions is:

$$y_j(t) = \sum_{i=1}^{r} y_i (t - 1) p_{ij} + u_j(t). \qquad (11.1.3)$$

If $\hat{p}_{ij}$ is an estimate of p_{ij}, then the predicted proportions are defined as[1]

$$\hat{y}_j(t) = \sum_{i=1}^{r} y_i (t - 1) \hat{p}_{ij}, \qquad j = 1, 2, ..., r. \qquad (11.1.4)$$

---

[1] We distinguish here between 'prediction' within the sample period and 'predic-
tion' outside the sample period. In the latter case, $\hat{y}_j(T_\alpha) = \sum_{i=1}^{r} \hat{y}_i(T_\alpha) \hat{p}_{ij}$,
$\alpha = T + 1, T + 2, ...$, both the $\hat{y}_i(T_\alpha)$ and $\hat{p}_{ij}$ are subject to error. This probably
means that as α gets larger, the variance of the error of prediction gets larger.

As an example of (11.1.4), when the data generated for the 1000 individuals are considered as an infinite population, the maximum likelihood estimates of the transition probabilities for the periods involved when $t = 3, 4, \ldots, 14$ are

$$[\hat{p}_{ij}] = \begin{bmatrix} 0.5613 & 0.3540 & 0.0847 & 0.0000 \\ 0.1212 & 0.5358 & 0.3430 & 0.0000 \\ 0.0000 & 0.0879 & 0.7209 & 0.1912 \\ 0.0000 & 0.0000 & 0.0861 & 0.9139 \end{bmatrix}. \qquad (11.1.5)$$

The observed and predicted proportions when (11.1.5) is used are given in table 11.1.

TABLE 11.1

The observed and predicted proportions for the simulated model (4.1.1) predicted by the ML estimator (11.1.5)

Time period	Observed proportions				Predicted proportions			
	s_1	s_2	s_3	s_4	s_1	s_2	s_3	s_4
2	0.4060	0.4280	0.1660	0.0000	—	—	—	—
3	0.2890	0.3850	0.2960	0.0300	0.2798	0.3877	0.3008	0.0317
4	0.2060	0.3250	0.3930	0.0760	0.2089	0.3346	0.3725	0.0840
5	0.1520	0.2900	0.4130	0.1450	0.1550	0.2816	0.4188	0.1446
6	0.1150	0.2620	0.4030	0.2200	0.1205	0.2455	0.4255	0.2115
7	0.0980	0.2170	0.3990	0.2860	0.0963	0.2165	0.4091	0.2781
8	0.0080	0.1700	0.3990	0.3430	0.0813	0.1861	0.3950	0.3377
9	0.0630	0.1750	0.3750	0.3870	0.0700	0.1573	0.3829	0.3898
10	0.0590	0.1430	0.3660	0.4320	0.0566	0.1491	0.3690	0.4254
11	0.0480	0.1180	0.3660	0.4680	0.0504	0.1297	0.3551	0.4648
12	0.0400	0.1130	0.3550	0.4920	0.0412	0.1124	0.3487	0.4977
13	0.0420	0.1040	0.3360	0.5180	0.0361	0.1059	0.3404	0.5175
14	0.0330	0.1050	0.3330	0.5290	0.0362	0.1001	0.3260	0.5377
15	—	—	—	—	0.0313	0.0972	0.3244	0.5471

To summarize the error of predictions, the mean square error may be used and for this problem is only 0.00066. A more sensitive measure of the error is the chi-square expression to be discussed in the next section.

11.2. The chi-square goodness-of-fit test

By the use of the asymptotic r variate normality of the multinomial distribution of the micro units $n_j(t)$, $j = 1, 2, ..., r$, and the fact that the quadratic form in the exponent of this distribution is distributed in the chi-square form with degrees of freedom equal to its rank $r - 1$ (Kendall and Stuart 1961, pp. 355–356), we have the Pearson tests of fit for the simple hypothesis that the observed and predicted proportions are identically distributed. The statistic which is a quadratic form is found to be

$$\chi^2_{(r-1)} = \sum_i (n_i(t) - N(t)\,\hat{y}_i(t))^2 / N(t)\,\hat{y}_i(t). \qquad (11.2.1)$$

The above expression may also be written as

$$\chi^2_{(r-1)} = \sum_i N(t)\,(y_i(t) - \hat{y}_i(t))^2 / \hat{y}_i(t). \qquad (11.2.2)$$

The expressions given by (11.2.1) or (11.2.2) are distributed as chi-square with $(r - 1)$ degrees of freedom and this forms the basis for the χ^2 goodness-of-fit test. Suppose that the proportions are predicted for T time periods. Then, according to the additive property of the chi-square distribution, the expression

$$\chi^2_{(r-1)T} = \sum_t^T \sum_i^r N(t)\,(y_i(t) - \hat{y}_i(t))^2 / \hat{y}_i(t) \qquad (11.2.3)$$

is chi-square distributed with $(r - 1)T$ degrees of freedom. The predicted proportion $\hat{y}_i(t)$ is assumed to be non-zero.

The statistic given by (11.2.3) can be employed to test whether the observed proportions or aggregate data are 'usual' or 'unusual' outcomes given that we assume that the observations have been generated by the relation (11.1.1), which is derived from a Markov process which assumes that the outcomes of the current period depend only on the outcomes of the immediately preceding period. If outcomes are deemed quite unusual on application of the χ^2 test, we shall *tentatively* conclude that the data may not be generated from the assumed process. We say tentatively because we realize that the occurrence of an unusual event is possible given that the assumed process is the one generating the data and also because a χ^2 test of the kind considered here does not stand in any simple relationship to measures of degree of confidence in the assumed model.

The chi-square value for the predicted proportions given in table 11.1 is 15.1998 which is far smaller than the tabled value 47.1420 with 36 degrees of freedom at the 10 percent significance level. Thus, on the basis of traditional sampling theory test procedures, the result is consistent with the hypothesis that the aggregate data are the outcomes of the Markov process given by (11.1.5).

We have shown in previous chapters that the ML, the GLS estimator and the MCS estimator all have the sample properties when the sample size is large. Recalling that the MCS estimator is obtained by minimizing the chi-square expression (7.2.2) given in ch. 7, one may think that the MCS estimator must have the minimum value of the chi-square value. Indeed, this is true for large samples because then the estimated weights which involve estimates of the true proportions are close to the true weights. This is to say that when the sample size is small, the MCS estimator may not have the minimum chi-square value. However, if we compute the modified chi-square value (7.3.2), the MCS estimator does have the smallest value among the alternative estimators. In the case of the large samples, expressions (7.2.2) and (7.3.2) are asymptotically equivalent.

11.3. Results from sampling experiment

More than 1500 problems have been processed for the 4 by 4 transition matrices that are estimated by the quadratic programming method. By the use of the IBM 7094, for each problem, the unrestricted and restricted estimates are computed, the predicted proportions are projected and finally the chi-square statistic is calculated for the test of goodness-of-fit purpose. None of the chi-square goodness-of-fit tests fail to accept the hypothesis that the data are generated by the Markov process. The means of the sample chi-square statistics for samples of size 25, 50 and 100 are given in table 11.2. The results of 1000 individuals are also listed showing the central tendency of the chi-square value.

TABLE 11.2
The chi-square value for sampling experiment

Estimator	Sample size			(Population 1000)	
	25	50	100	χ^2	Modified
Restricted classical least squares	14.2943	17.6977	17.6358	15.9293	16.1164
Restricted weighted least squares (the weights are the inverse of mean proportions)	14.2899	17.6459	17.5826	16.0009	15.1996
ML (GLS, MCS) estimator	16.1373	19.5361	18.4240	15.1998	15.1526
Bayesian estimator with prior means approximately equal to true parameter values	19.0119	22.1620	21.2785	—	—

The modified chi-square values are also computed for the 1000 population. The ML estimator does have the smallest modified chi-square value. The classical estimator seems to be worse in terms of this measure. All the computed chi-square values in table 11.2 are smaller than the tabled value for 36 degrees of freedom at 10 percent significant level, namely 47.14.

CHAPTER 12

COMPARISONS OF THE ALTERNATIVE ESTIMATORS

In the previous chapters, alternative estimators have been introduced, and the sampling experiments have been presented for each estimator. Now, a summary of the experimental results will be given and the performance of the alternative estimators compared.

12.1. The basis for comparison

In order to make a valid comparison, the same set of data must be used in the computation of all estimates. In previous experiments, two sets of data were used. The original set of data used periods 1 to 16 and the other set of data made use of periods 3 to 14. To make a comparison of all the estimators, the set of data for periods 3 to 14 is used. Therefore, the unweighted classical estimates and the estimates weighted by the inverse of the mean proportions are recomputed for this purpose. The means and the root mean square error from true proportions for the unrestricted classical estimates, restricted classical estimates and the restricted estimates weighted by the inverse of the mean proportions are given in tables 12.1, 12.2 and 12.3.

12.2. Aggregate mean square error and variance measure

A question arises as to how to compare the root mean square error matrices of the alternative estimators when the elements of one matrix are not unanimously larger or smaller than those of another matrix. One criterion which may be used is to compare the aggregate mean square errors which are the unweighted sum of the elements of the root mean square error matrices. This measure involves equal weighting of

145

TABLE 12.1

The unrestricted classical estimates from the amended data
(sample periods 3 to 14)

Sample size	Means				Root mean square error			
25	0.4416	0.5808	0.1231	−0.1455	0.2611	0.3534	0.4454	0.4007
	0.1605	0.3631	0.4439	0.0326	0.1522	0.2422	0.3022	0.2533
	0.0117	0.1388	0.5398	0.3097	0.1360	0.1515	0.2407	0.1747
	0.0064	−0.0201	0.2151	0.7986	0.0938	0.1068	0.2054	0.1635
50	0.5041	0.5134	0.0167	−0.0342	0.1668	0.2680	0.2816	0.2027
	0.1544	0.3715	0.5155	−0.0414	0.1323	0.2208	0.2652	0.1989
	−0.0127	0.1795	0.5284	0.3047	0.1220	0.1861	0.2343	0.1664
	0.0147	−0.0514	0.2294	0.8073	0.0888	0.1299	0.1808	0.1376
100	0.5161	0.5494	−0.0063	−0.0591	0.1605	0.2670	0.3039	0.2135
	0.8151	0.2913	0.5166	0.0071	0.1600	0.2599	0.2960	0.2160
	−0.0300	0.2107	0.5787	0.2406	0.0885	0.1492	0.1689	0.1067
	0.0109	−0.0513	0.1748	0.8656	0.0569	0.1015	0.1165	0.0766

TABLE 12.2

Restricted classical estimates from the amended data (sample periods 3 to 14)

Sample size	Means				Root mean square error			
25	0.3756	0.4788	0.1400	0.0056	0.2914	0.2408	0.2445	0.0203
	0.1628	0.4242	0.3957	0.0172	0.1295	0.1906	0.1682	0.0385
	0.0377	0.1071	0.5949	0.2604	0.0613	0.0796	0.1679	0.1096
	0.0082	0.0130	0.1687	0.8100	0.0195	0.0288	0.1244	0.1392
50	0.4721	0.4587	0.0677	0.0015	0.1766	0.1815	0.1255	0.0061
	0.1389	0.4409	0.4112	0.0091	0.1017	0.1577	0.1186	0.0241
	0.0168	0.1077	0.6242	0.2513	0.0329	0.0705	0.1286	0.0827
	0.0034	0.0147	0.1599	0.8220	0.0084	0.0319	0.0985	0.1081
100	0.4947	0.4268	0.0785	0.0000	0.1520	0.1685	0.1336	0.0000
	0.1553	0.4452	0.3932	0.0063	0.0964	0.1481	0.1064	0.0149
	0.0069	0.1151	0.6608	0.2172	0.0135	0.0453	0.0730	0.0412
	0.0009	0.0063	0.1222	0.8706	0.0028	0.0177	0.0518	0.0539

TABLE 12.3

Restricted estimates weighted by the inverse of the mean proportions
from the amended data (sample periods 3 to 14)

Sample size	Means				Root mean square error			
25	0.4123	0.4660	0.1157	0.0052	0.2598	0.2206	0.1997	0.0196
	0.1509	0.4338	0.3989	0.0184	0.1163	0.1862	0.1482	0.0403
	0.0304	0.1061	0.6058	0.2577	0.0546	0.0811	0.1589	0.1081
	0.0120	0.0144	0.1607	0.8132	0.0249	0.0313	0.1265	0.1335
50	0.4985	0.4442	0.0559	0.0016	0.1557	0.1667	0.1090	0.0066
	0.1295	0.4546	0.4065	0.0093	0.0969	0.1484	0.1102	0.0244
	0.0137	0.1038	0.6349	0.2476	0.0301	0.0698	0.1207	0.0801
	0.0055	0.0153	0.1525	0.8268	0.0122	0.0325	0.0925	0.1039
100	0.5216	0.4162	0.0622	0.0000	0.1335	0.1504	0.1098	0.0000
	0.1417	0.4537	0.3578	0.0068	0.0887	0.1326	0.0917	0.0162
	0.0056	0.1132	0.6672	0.2142	0.0115	0.0453	0.0669	0.0400
	0.0020	0.0073	0.1165	0.8742	0.0052	0.0195	0.0486	0.0509

each parameter and does not take into account dependencies among the errors associated with estimates of the p_{ij}'s.

The aggregate mean square error of the results presented in tables 4.2, 4.5 and 4.6 of ch. 4, tables 5.1, 5.2 and 5.3 of ch. 5 and table 10.3 of ch. 10 are listed in table 12.4. This group is based on the sample data from periods 1 to 16.

TABLE 12.4

Sum of the elements of the root mean square error matrices
from some estimators (sample periods from 1 to 16)

Sample	Micro Maximum likelihood	Unrestricted least squares	Restricted minimum absolute deviations	Restricted least squares			
				Unweighted	Weighted by proportion means	Estimate of Disturbance variance	Product of sample proportions
25	0.4025	2.4944	1.5014	1.4795	1.4052	1.3812	1.4243
50	0.3155	2.0537	1.1978	1.0191	0.9915	0.9715	0.9981
100	0.2190	1.9526	1.0453	0.8520	0.7931	0.8002	0.8108

From table 12.4, it appears that by making use of the aggregate root mean square error criterion, although the choice among the weights employed is arbitrary, the weighted restricted least squares estimator has a smaller aggregate RMSE than the other least squares and minimum absolute deviations estimators employed. Furthermore, there appears to be a great gain in efficiency as measured by the aggregate RMSE, in going from the unrestricted to the restricted least squares estimators. However, if the micro data as opposed to the proportion data are available, then, of course, the micro maximum likelihood estimator is greatly superior, on a RMSE basis, to any of the macro estimators employed in this study. Of course, as noted in ch. 2, it is also possible to compute Bayesian estimates based on the micro data.

The aggregate mean square errors of the results presented in tables 6.2 and 6.3 of ch. 6, tables 9.5, 9.6, 9.11 and 9.12 in ch. 9 and tables 12.1, 12.2 and 12.3 of this chapter are listed in table 12.5. This set of results is based on the sample data for periods 3 to 14.

TABLE 12.5

Aggregate root mean square errors for alternative estimators
(sample periods 3 to 14)

| Estimator | Sample size | | | | | |
| | 25 | | 50 | | 100 | |
	Unrest.	Rest.	Unrest.	Rest.	Unrest.	Rest.
Restricted classical least squares	3.6829	2.0641	2.9920	1.4536	2.7418	1.1191
Restricted weighted least squares	3.6829	1.9095	2.9920	1.3597	2.7418	1.0109
Maximum likelihood (or GLS, MCS)	3.5181	1.7440	2.9800	1.2726	2.7936	0.9750
Bayesian with platy-kurtic prior and equal prior means	2.3793	1.5973	2.2678	1.2003	1.9659	0.9167
Bayesian with lepto-kurtic prior and prior means approximately equal to true parameter values	0.0383	0.0265	0.0500	0.0415	0.0503	0.0391

Generally speaking, using the root mean square criterion, the ML (GLS, MCS) estimator is better than the unrestricted least squares estimator and also the estimator weighted by the inverse of the observed proportions. The Bayesian estimator has a smaller aggregate root mean square error for each sample size than the ML estimator. The degree of improvement depends on the parameters of the multi-beta prior distribution used. The results show that if we have prior knowledge about the values of the parameters, use of this knowledge in estimation results in a considerable reduction in RMSE's.

To compare the variances of the alternative estimators, the aggregate sample variances, the sum of the sample variances of the distribution of the estimates for each parameter have been computed. They are given in table 12.6.

TABLE 12.6

Aggregate variances for alternative estimators (sample periods for 3 to 14)

| Estimator | Sample size | | | | | |
| | 25 | | 50 | | 100 | |
	Unrest.	Rest.	Unrest.	Rest.	Unrest.	Rest.
Restricted classical least squares	0.8255	0.2597	0.4740	0.1334	0.4290	0.0997
Restricted weighted least squares	0.8255	0.2252	0.4740	0.1217	0.4290	0.0836
Maximum likelihood (or GLS, MCS)	0.8047	0.2085	0.4938	0.1237	0.4288	0.0832
Bayesian with platy-kurtic prior and equal prior means	0.2379	0.1211	0.2178	0.0859	0.1694	0.0557
Bayesian with lepto-kurtic prior and prior means approximately equal to true parameter values	0.0001	0.0000	0.0001	0.0000	0.0001	0.0000

Among the non-Bayesian estimators, the ML (GLS, MCS) is the best in the sense that it has the smallest variance. The use of prior information associated with the Bayesian estimators has resulted in a considerable reduction in variance relative to those associated with other estimators. When the leptokurtic prior is used in the present experiments, there is a large variance reduction. On the other hand, when the platykurtic prior is used, the variance of the Bayesian estimator is reduced relative to that of the ML estimator but the reduction is not as great as in the case of the Bayesian estimator based on the leptokurtic prior.

12.3. Wilcoxon matched-pairs signed-ranks test and Kendall's coefficient of concordance

In addition to the above criteria for measuring performance of alternative estimators, the Wilcoxon matched-pairs signed-ranks test (Siegel 1956, pp. 75–83) and a test of the rank association employing Kendall's coefficient of concordance (Siegel 1946, pp. 229–239) have been applied. The former is designed to test the hypothesis that the absolute deviations from the true parameter value of a pair of estimators are the same. The latter provides a means of testing the hypothesis that there exists no rank association among alternative estimators.

Of the results to be reported, fifty estimates of sample size 50 are used for each estimator and two groups of comparisons are made. In

TABLE 12.7
The Wilcoxon matched-pair

	p_{11}	p_{12}	p_{13}	p_{14}	p_{21}	p_{22}	p_{23}
RLSWM RLSC	-1.61	-1.08	-2.28	-1.00	0.191	-1.62	-2.93
RLSWM MAD	-0.95	-0.61	-0.53	-1.00	0.03	-1.20	-1.05
RLSWM URLSC	-2.01	-3.20	-5.49	-6.15	-3.73	-3.91	-4.84
RLSC MAD	-0.80	-0.58	-0.19	-1.00	-0.32	-1.98	-0.77
RLSC URLSC	-1.46	-3.13	-5.53	-6.15	-3.62	-3.82	-4.78
MAD URLSC	-1.11	-1.75	-4.62	-6.15	-3.10	-1.63	-3.47

the first group, all the estimates are obtained from the original experimental data and the comparisons are made among the following four alternative estimators: restricted least squares weighted by mean (RLSWM), restricted least squares classical (RLSC), minimum absolute deviations (MAD), and unrestricted least squares classical (URLSC) estimators. These results are given in tables 12.7 and 12.8. In the second group, all the estimates are obtained from the amended data in which no zero proportion is involved. The comparisons are made among URLSC, RLSWM, ML, URML and two types of restricted Bayesian estimators; Bayesian with platykurtic prior and equal prior means (in the table, called Bayesian II) and Bayesian with leptokurtic prior and prior means (called Bayesian I) equal to the true parameter values. These comparisons are given in table 12.10. The results from two groups of comparisons are consistent.

The Wilcoxon matched-pair signed-rank tests show that there are significant differences between the restricted and unrestricted estimators at the five percent level of significance. All of the statistical tests are significant except for p_{11} and p_{44}. When we compare the RLSWM estimator and the RLSC estimator, 12 out of 16 estimated parameters are in favor of the RLSWM estimator, although most of them are not significantly better even at the ten percent level of significance (table 12.7).

signed-ranks test

p_{24}	p_{31}	p_{32}	p_{33}	p_{34}	p_{41}	p_{42}	p_{43}	p_{44}
−0.50	−3.06		−1.70	−2.95			−0.98	−1.77
		−1.14			−3.38	−1.14		
−2.86	−0.82	−0.78	−0.99	−2.01	0.79	−1.14	−1.26	−2.99
−6.15	−5.94	−5.10	−4.11	−6.14	−5.76	−3.65	−3.65	−1.35
−2.86	−1.23	−0.69	−1.23	−2.16	0.10	−1.06	−1.31	−2.20
−6.15	−5.79	−5.20	−4.28	−3.69	−6.15	−5.75	−3.77	−1.19
−6.15	−6.15	−4.49	−3.63	−4.60	−5.89	−5.21	−3.15	−2.84

When we compare the RLSWM estimator and the MAD estimator, most of the elements show no significant difference except p_{24} which is in favour of MAD. In general, the MAD estimator estimates zero parameter such as p_{14}, p_{41}, etc. very well while the RLSWM estimator gives better

TABLE 12.8

Ranking and the Kendall coefficient of concordance

Parameters		Ranking				Sum of square deviations	Coefficient of concordance
True value	Estimate	RLSWM	RLSC	MAD	URLSC		
0.6	p_{11}	113.5	119.5	131.0	116.0	319.5	0.0256
0.4	p_{12}	110.5	117.5	123.0	149.0	846.5	0.0679
0	p_{13}	99.0	108.5	113.5	179.0	3996.5	0.3824
0	p_{14}	100.0	101.0	99.0	200.0	7502.0	0.9871
0.1	p_{21}	113.0	113.0	122.0	152.0	1026.0	0.0827
0.5	p_{22}	111.0	116.0	124.0	149.0	854.0	0.0683
0.4	p_{23}	97.0	118.0	122.0	163.0	2286.0	0.1829
0	p_{24}	105.5	107.5	87.0	200.0	7755.5	0.8499
0	p_{31}	99.5	113.5	97.0	190.0	5791.5	0.6178
0.1	p_{32}	114.5	109.0	110.0	166.5	2313.5	0.1869
0.7	p_{33}	107.0	118.5	109.0	165.5	2262.5	0.1817
0.2	p_{34}	106.5	130.5	103.0	160.0	2081.0	0.1675
0	p_{41}	107.0	92.5	104.5	196.0	6841.5	0.6893
0	p_{42}	108.5	100.5	105.0	186.0	4993.5	0.5135
0.1	p_{43}	115.5	118.5	114.0	152.0	982.5	0.0788
0.9	p_{44}	120.5	135.5	108.0	136.0	540.5	0.0435

estimates for non-zero true parameters. When we compare the RLSC estimator with the MAD estimator, the situation is similar to that of the relation between the RLSWM and MAD estimators and most of them show no significant difference. The numbers of estimators showing significant superiority are given in table 12.9.

The Kendall coefficient of concordance which expresses the degree of association among the ranking of the alternative estimators is computed and the results are given in table 12.8. The procedure is as follows: the same set of data is used to compute the four alternative estimators (RLSWM, RLSC, MAD and URLSC). The absolute deviations of the estimators from the true value are compared and the ranks 1, 2, 3 and 4 are assigned to the four estimators; 1 for the closest estimator and 2 for the next closest estimator and so on. If two estimators are identical, the average rank (e.g., 1.5 or 2.5, etc.) is assigned to each estimator to break the tie. Each parameter is judged separately. Fifty

TABLE 12.9

Number of p_{ij} for which estimator r was superior to estimator s by Wilcoxon test at 5 percent significance level (sample periods 1 to 16)

r	s: RLSWM	RLSE	MAD	URLSC
RLSWM	—	6	0	15
RLSC	2	—	1	14
MAD	2	3	—	14
URLSC	0	0	0	—

sets of data are used and thus 50 sets of rankings are obtained for each parameter. The sum of 50 rankings is then computed for each estimator and is listed in table 12.8. If the 50 sets of rankings are in perfect agreement about a best estimator, then one estimator would have received 50 ranks of 1 and thus its sum of ranks would be 50. It should be clear

TABLE 12.10

Out of 16 elements, number of p_{ij} for which estimator r was superior to estimator s by Wilcoxon test at 5 percent significance level (sample periods 3 to 14)

r	s: Bayes I	Bayes II	ML	RLSW	RLS	UR Bayes I	UR ML	UR RLS
Bayesian I	—	14	14	14	14	9	16	16
Bayesian II	0	—	7	9	10	—	—	—
ML	0	3	—	3	6	2	12	15
RLSW	1	3	2	—	10	2	15	15
RLS	1	2	1	1	—	2	14	15
UR Bayesian I	2	—	14	14	14	—	16	16
UR ML	0	—	3	0	0	0	—	2
UR RLS	0	—	0	0	0	0	1	—

that the degree of agreement among the 50 sets of rankings is reflected by the degree of variance among the four sums of ranks. The coefficient of concordance is a function of that degree of variance. It is computed as a ratio of the variance of ranks and the maximum possible variance which would occur with perfect agreement among 50 rankings.[1] Thus, the coefficient of concordance is between zero and one. If the four estimators had the same sum of ranks, the coefficient of concordance would be zero. If the 50 rankings are in perfect agreement, the coefficients of concordance would be one. From table 12.8, Kendall coefficients of concordance, the W's, are not high except for the estimates of $p_{14}, p_{24}, p_{31}, p_{41}$ and p_{42} whose true parameter values are all zero.

Examining the sum of the ranks, we find that differences among the restricted estimators are not significantly different but the restricted estimators are all superior to the unrestricted classical least squares estimators. Using p_{14} as an example, the ranks are 100, 101 and 99 for the RLSWM, RLSE and MAD estimators respectively, but the URLSC estimator is ranked 200, which means the worst. The test about S, the sum of the squared deviation of the ranks from their means, by chi-square also shows that most of the dispersions are not large enough to be considered as significant differences in rank.

From the second group comparisons, as reflected in table 12.10, the unrestricted estimates are not acceptable and have the following ranks:

---

[1] If there is no tie in rankings, the formula used in computing W is

$$W = \frac{s}{k^2 (N^3 - N)/12},$$

where
s = sum of squares of the observed deviations from the mean of the sum of ranks
k = 50 sets of rankings,
N = 4 estimators.

If there are ties in rankings, the correction factor should be computed as

$$T = \Sigma (t^3 - t)/12,$$

where t is the number of observations in a group tied for a given rank. The summation directs one to sum over all groups of ties within any one of 50 rankings. The Kendall coefficient of concordance is

$$W = \frac{s}{k^2 (N^3 - N)/12 - k\Sigma_T T},$$

where $\Sigma_T T$ directs one to sum the value of T for all the k rankings (Siegel 1956).

unrestricted Bayesian, unrestricted maximum likelihood, and unrestricted least squares classical estimators. The unrestricted Bayesian is much better than the URMLE and the URLSC. However, the supremacy of the URMLE to the URLSC is not very significant. Among the restricted estimators, the Bayesian estimator is the best. The ML estimators are better than the RLSWM (or RLSW) estimates in the sense that for estimation of 11 of the 16 transition probabilities, they are closer to the true parameter values. The results of the Wilcoxon test indicate that only 3 transition probabilities show significant difference at the 5 percent significance level. The RLSWM is also better than the RLSC and not more than one-half of the elements show significant differences. The unrestricted Bayesian estimator is not acceptable (in the sense that some of the estimates are negative) but it is quite close, in terms of the absolute deviation, to the true parameter values; closer in fact than the other non-Bayesian estimators. The Bayesian estimates with equal prior means and large prior variance (Bayesian II) are, of course, not as good as the Bayesian estimates with true prior means and small prior variances (Bayesian I), but are better than the ML, the RLSWM and also RLSC in that most of the estimates are close to the true parameter values.

The Kendall coefficient of concordance computed for all the alternative estimators has a range from 0.5 to 0.8 and thus indicates a high degree of association in the ranking. The results also show that the following ranking is significant: the Bayesian and the unrestricted Bayesian rank 1, the MLE, the RLSWM and the RLSC estimators rank 2 and the URML and the URLSC estimators rank 3. In each group of the same rank, the estimators may be ranked again, however, those within the ranks are not significant.

In conclusion, the rankings based on the pair-wise comparison among restricted estimators are, in general, similar to those based on the root mean square error criterion. However, the results of the tests for all parameters are not unanimous and thus final, invariant conclusions are not possible. One thing that stands out is the superiority of the restricted over the unrestricted estimators and the gain in efficiency when the maximum likelihood and Bayesian estimators are used.

12.4. Summary

In summary, the results of this experiment are:

(1) In terms of a RMSE criterion, the ML estimator from the micro data is superior to any other estimator using aggregate data (table 12.4). Among the estimators using aggregate data, the restricted estimator is better than the unrestricted estimator, the restricted least squares estimators are better than the restricted minimum absolute deviations estimator, the weighted restricted estimator is better than the unweighted restricted estimator. Among the weighted estimator, the ML (or GLS, MCS) estimator is the best. If prior knowledge is available, the Bayesian estimator with leptokurtic prior and prior means approximately equal to true parameter values has a smaller RMSE than any other estimator.

(2) With respect to the variances of alternative estimators, the estimators may be ranked in the same order as (1) above. The variances of the estimators are reduced as one goes from the classical, weighted, maximum likelihood to the Bayesian estimator. Correct prior knowledge in the Bayesian context substantially reduces the variance of the estimator (table 12.6).

(3) On the bias of alternative estimators, it was found that when sample size increases, the bias of estimators tends to vanish.

(4) The distribution of the estimators tended to normality fairly rapidly. For example, the Kolmogorov–Smirnov normality test performed in ch. 6 showed that none of the estimates whose true parameter values are nonzero is significantly different from the normal distribution at the 10 percent significance level.

(5) Results on the goodness-of-fit test showed that none of the tests succeeds in rejecting the hypothesis that the data are generated by the specified Markov process. Since the observed proportions were used to substitute for the true proportions in estimating the weight matrix, the modified minimum chi-square estimator does have the minimum modified chi-square value.

CHAPTER 13

CONCLUDING REMARKS

In the introductory chapter, we asked whether it was possible to use aggregate outcome (proportion) data to estimate transition probabilities associated with a first order Markov model assumed to represent the behavior of micro units. In view of the analysis and results presented in the preceding chapters, this question can now be answered affirmatively.

As the starting point in the search for suitable estimators, unrestricted and restricted least squares estimators were presented and their sampling properties were investigated in Monte Carlo experiments. The results of the experiments confirmed theoretical results by showing that in large samples, the estimates converged to the true parameter values and tended to be normally distributed. Also, it was found that on the basis of an MSE criterion, the restricted estimator performed better than the unrestricted estimator. And, of course, use of the restricted estimator insures that estimates will fall in the admissible region of the parameter space.

Next, we formulated several weighted restricted least squares estimators. Weights were introduced in an effort to mitigate the effects of heteroscedasticity. The Monte Carlo experimental results revealed that the weighted least squares estimators have smaller MSE's than do unweighted least squares estimators and retain the properties of consistency and large sample normality. These experimental results thus conform to theoretical expectations.

To circumvent the problem associated with the singularity of the disturbance covariance matrix, we found that we could delete one stochastic relation from the system without any loss of information to obtain a 'reduced' system. Both restricted and unrestricted estimators for the parameters of the reduced system, which take account of heteroscedasticity, were generated by an application of a generalized least squares approach. Parameters in the deleted equation were then estimated by use of the 'adding up' constraint on the parameter vectors. It was found that estimators so obtained were unique no matter which equation of the system was deleted to obtain the reduced system. The results of Monte Carlo experiments showed that the estimator yielded by the generalized least squares approach performed better than any of those mentioned above. The properties of consistency and large sample normality for the generalized best squares estimator were confirmed by the experimental results.

To generate yet another estimator, the minimum chi-square (MCS) approach was employed. For known disturbance covariance matrix, it was shown that the MCS estimator is identical to the GLS estimator. When the unknown disturbance covariance matrix appearing in the MCS estimator is replaced by a consistent estimate, we obtain an approximation to the MCS which we call the minimum modified chi-square (MMCS) estimator. The MCS and MMCS estimators are both best asymptotic normal (BAN) estimators.

In pursuing a maximum likelihood (ML) approach to obtain an estimator based on the aggregate proportion data, it was found that the ML estimator had exactly the same form as the MCS and GLS estimators. In fact, the ML and the MCS methods are dual problems of finding a saddle point in which in one case, one obtains a maximum from below and in the other, its minimum from above. An iterative feed back procedure which utilizes successive estimates of the transition probabilities to estimate parameters of the disturbance covariance matrix was introduced. Since restrictions are in the form of inequalities, this is a recursive quadratic programming problem. Experience with the recursive solutions indicates that convergence is obtained in three or four iterations. Solutions will not explode since the estimates are in the closed interval, zero to one.

In the Bayesian approach to the problem of analyzing the first order Markov process with aggregate proportion data, we employed a multivariate beta prior distribution for the parameters. This prior distribution incorporates the information that transition probabilities are defined for the closed interval zero to one and also that they satisfy 'adding up' constraints. Further by choice of prior parameters, an investigator can introduce prior information about the parameters if such information is available. Of course, if a good deal of prior information is combined with the likelihood function, it will have a decided influence on the location, spread and shape of the posterior distribution in small and moderate sized samples. On the other hand, if the prior distribution is relatively flat over the region in which the likelihood function assumes appreciable values, then the posterior distribution's features will be largely determined by the shape of the likelihood function. In either case, as the sample size grows large, the posterior distribution's properties will be determined by the form of the likelihood function, as is well known. Thus, *in large samples*, the posterior distribution will be well approximated by the likelihood function and will have a multivariate normal form with the mean approximately equal to the ML estimate.

Since many have an interest in the sampling properties of estimators, sampling experiments were performed to assess the sampling properties of two Bayesian estimators. One was generated using a platykurtic prior distribution while the other was based on a leptokurtic prior distribution. The former prior is less informative about the parameters than is the latter. Results of the sampling experiments showed that the Bayesian estimators considered performed better than all other estimators. This was particularly the case when a leptokurtic prior distribution was employed. Even when a platykurtic distribution was employed, results for the Bayesian estimator were better than for the ML (GLS, MCS) estimator. On the basis of these experimental results and other more fundamental considerations, we find much merit in the Bayesian approach to the analysis of the model and the results yielded by it.

In quite a different approach, an estimation procedure which minimizes the sum of absolute deviations (MAD), was developed. The non-negativity restrictions and the row sum constraints for the transition probabilities are imposed by use of a conventional linear programming method. The results of the sampling experiments show that estimates so obtained converge to true parameter values and are normally distributed in large samples. However, the MAD estimator is not admissible in the sense that this estimator has a larger MSE than those associated with other estimators.

Prediction and chi-square tests of goodness-of-fit indicate that although the unrestricted estimates may give good predictions, sometimes the predicted proportions are negative and thus meaningless. The closeness of the prediction, as measured by the chi-square value, may be used as a basis for testing the hypothesis that the data are generated from a Markov model. When this test is employed, the results are consistent with the hypothesis that the data are generated from a Markov process.

In summarizing the sampling experiments, a comparison of the alternative estimators is made by means of the aggregate mean square error criterion. The experimental results indicate that the ML (GLS, MCS) estimator is the best among the non-Bayesian estimators. The Bayesian estimators considered performed better than all other estimators, in part because they take advantage of both sample and prior information. The variances for the estimators are also compared. The conclusion is the same as that obtained via the mean square error route.

The Wilcoxon matched-pair signed-ranks tests also give results similar to those based on the root mean square error criterion. Although the test results are not entirely consistent, the alternate criteria employed in ascertaining the relative performance of alternative estimators suggest the following ranking: Bayesian, maximum likelihood, weighted restricted least squares, unweighted restricted least squares, minimum absolute deviations, and unrestricted least squares estimators.

In order to demonstrate application of the procedures for estimating transition probabilities in economic problems, the Telser problem, concerned with the behavior of cigarette smokers, and the problem of tenure status in Taiwan are studied. The results are consistent with observed economic behavior and indicate that estimation procedures developed herein yield reasonable results.

In the past, one of the major problems in making use of the Markov probability model has been the necessity of having detailed time ordered data. Since the techniques evaluated in this book are operational, they should go a long way toward reducing or eliminating the data problem of the past and should open the way to a variety of applications where the data are thought to be generated in a way consistent with a first order Markov process.

It should be noted that all the work presented in this book is based on the important assumption that the transition probabilities are constant over the sampling periods (stationarity). This may not always be true since behavior may not be time invariant especially when an economy may be disturbed by exogenous factors such as wars, new economic policies, etc. Therefore, the transition probabilities may change through time and the so-called non-stationary Markov chain results. The problem remaining is how to trace the changes of the transition probabilities when the micro observations are not available and the data one may work with are aggregate in nature. One way of coping with this estimation problem is given in appendix C.

THE GENERALIZED INVERSE METHOD

As indicated in ch. 5, Aitken's generalized least squares method cannot be employed in the case of the transition probability regression model because the corresponding disturbance covariance matrix is singular. In this situation, the generalized inverse method has been suggested as one way to solve this type of problem. Within this context, the purpose of this appendix is to investigate the use of the generalized inverse method of regression in estimating the transition probabilities.

A.1. A generalization of the generalized least squares

In this section, the notation will be initally defined in a very general form. Let the linear model be

$$y = Xp + u, \qquad (A.1.1)$$

where y is a $(T \times 1)$ vector and X is a $(T \times k)$ matrix of observations, p is a $(k \times 1)$ vector of parameters and u is a $(T \times 1)$ vector of disturbances with assumptions

$$Eu = 0, \qquad (A.1.2)$$

$$Euu' = \Sigma, \qquad (A.1.3)$$

and Σ is singular of rank g which is smaller than the number of total observations T[1].

If Σ were nonsingular, the most efficient way to estimate p is to transform the observations in (A.1.1) such that the resulting disturb-

---

[1] In our special Markov chain regression case, y is a $(rT \times 1)$ vector, X is a $(rT \times r^2)$ matrix, p is a $(r^2 \times 1)$ vector and u a $(rT \times 1)$ vector, and Σ has rank $(r-1)T$ which is less than the full rank rT.

ances are homoscedastic and then use the least squares estimator. The transformation may be done by multiplying a $(T \times T)$ nonsingular matrix H' with relation (A.1.1),

$$H'y = H'Xp + H'u \qquad \text{(A.1.4)}$$

so that

$$EH'u = 0 \qquad \text{(A.1.5)}$$

and

$$EH'uu'H = H'\Sigma H = I_T, \qquad \text{(A.1.6)}$$

where I_T is a $(T \times T)$ identity matrix. The result of applying least squares to (A.1.4) is that

$$\bar{p} = (X'\Sigma^{-1}X)^{-1}X'\Sigma^{-1}y, \qquad \text{(A.1.7)}$$

where

$$\Sigma^{-1} = HH'. \qquad \text{(A.1.8)}$$

Since our disturbance covariance matrix Σ is singular, Σ^{-1} does not exist and thus there exists no such H matrix.

If Σ is singular and of rank g, we propose to find a $(T \times g)$ matrix K of rank g which transforms (A.1.1) into

$$K'y = K'Xp + K'u, \qquad \text{(A.1.9)}$$

such that

$$EK'u = 0, \qquad \text{(A.1.10)}$$

and

$$EK'uu'K = K'\Sigma K = I_g, \qquad \text{(A.1.11)}$$

where I_g is a $(g \times g)$ identity matrix.[2] Thus K is a generalization of H, since if Σ has full rank then K is equal to H. After application of the least squares criterion to (A.1.9), we have

$$\bar{p} = (X'\Sigma^+X)^{-1}X'\Sigma^+y, \qquad \text{(A.1.12)}$$

where

$$\Sigma^+ = KK'. \qquad \text{(A.1.13)}$$

---

[2] It should be noted that T observations are transformed into g observations. Thus, the question arises as to whether the estimator based on (A.1.9) is efficient. See also footnote 4.

Σ^+ is the generalized inverse of Σ, and has the following properties:[3]

$$\Sigma\Sigma^+\Sigma = \Sigma, \tag{A.1.14}$$

$$\Sigma^+\Sigma\Sigma^+ = \Sigma^+, \tag{A.1.15}$$

$$(\Sigma\Sigma^+)' = \Sigma\Sigma^+, \tag{A.1.16}$$

$$(\Sigma^+\Sigma)' = \Sigma^+\Sigma, \tag{A.1.17}$$

In the following, we will prove the existence of K and that (A.1.12) and (A.1.13) hold.

We will start with the singular matrix Σ. Since Σ is a real symmetric matrix, there exists an orthogonal matrix U, the matrix of characteristic vectors, such that $U'\Sigma U$ is a diagonal matrix whose diagonal elements are the characteristic roots of Σ, i.e.,

$$U'\Sigma U = \Lambda. \tag{A.1.18}$$

If the matrix Σ is singular and of rank g, there must be a $(T-g)$ characteristic roots with values zero. Thus,

$$\Lambda = \begin{bmatrix} \lambda_1 & & & & & & \\ & \lambda_2 & & & & & \\ & & \ddots & & & & \\ & & & \lambda_g & & & \\ & & & & 0 & & \\ & & & & & \ddots & \\ & & & & & & 0 \end{bmatrix} = \begin{bmatrix} \lambda & \vdots & \\ \cdots & \vdots & \cdots \\ & \vdots & 0 \end{bmatrix}, \tag{A.1.19}$$

where the λ_i's may be arranged in descending order, $\lambda_1 \geq \lambda_2 \geq \lambda_3 \cdots \geq \lambda_g$. Thus, it is possible to obtain

$$\Lambda^{-\frac{1}{2}}U'\Sigma U\Lambda^{-\frac{1}{2}} = \begin{bmatrix} I_g & \\ & 0 \end{bmatrix}, \tag{A.1.20}$$

[3] For a discussion of the concept of the generalized inverse, see Boot (1963), Penrose (1955), Rado (1956), Rao (1962) and Price (1964).

where $\Lambda^{-\frac{1}{2}}$ is a diagonal matrix with elements which are the reciprocal of the square root of Λ_i, and the last $T - g$ elements are zeroes. Let the orthogonal matrix be partitioned into

$$U = (U_1 \vdots U_2), \qquad (A.1.21)$$

where U_1 has g columns and U_2 has $T - g$ columns. Thus we have

$$U\Lambda^{-\frac{1}{2}} = (U_1 \vdots U_2) \begin{pmatrix} \lambda^{-\frac{1}{2}} & \\ \cline{1-2} & 0 \end{pmatrix} = (U_1\lambda^{-\frac{1}{2}} \vdots 0) \quad (A.1.22)$$

where $U_1\lambda^{-\frac{1}{2}}$ is a $(T \times g)$ matrix which we will call K, i.e.,

$$K = U_1\lambda^{-\frac{1}{2}}. \qquad (A.1.23)$$

Hence, we have

$$K'\Sigma K = \lambda^{-\frac{1}{2}}U_1\Sigma U_1\lambda^{-\frac{1}{2}} = I_g, \qquad (A.1.24)$$

which proves the existence of K.

Next, from (A.1.23), we have

$$U_1 = K\lambda^{\frac{1}{2}}, \qquad (A.1.25)$$

where $\lambda^{\frac{1}{2}}$ is a $(g \times g)$ diagonal matrix with elements which are the square root of λ_i. Since U has T orthogonal vectors, $U^{-1} = U'$. Therefore, the covariance matrix Σ may be written as

$$\Sigma = U\Lambda U' = (U_1 \vdots U_2) \begin{pmatrix} \lambda & \\ \cline{1-2} & 0 \end{pmatrix} \begin{pmatrix} U_1' \\ \cline{1-1} U_2' \end{pmatrix} = U_1\lambda U_1', \quad (A.1.26)$$

which by the use of (A.1.25) becomes

$$\Sigma = U_1\lambda U_1' = K\lambda^{\frac{1}{2}}\lambda\lambda^{\frac{1}{2}}K' = K\lambda^2 K' = (K\lambda)(K\lambda)'. \quad (A.1.27)$$

According to Deutsch (1965, p. 83), if a singular matrix Σ can be factored into products of nonsquared matrices $(K\lambda)$ with full rank, then by the use of (A.1.23)

$$K'K\lambda = \lambda^{-\frac{1}{2}}U_1'U_1\lambda'^{-\frac{1}{2}}\lambda = I_g, \qquad (A.1.23a)$$

and the generalized inverse is given by

$$\Sigma^+ = (K\lambda)(\lambda K'K\lambda)^{-1}(\lambda K'K\lambda)^{-1}(K\lambda)' = KK', \qquad (A.1.28)$$

which completes the proof of (A.1.13).

Given these results, if we make use of the least squares estimator for (A.1.9), under certain conditions[4] we obtain the best linear unbiased estimate of $\bar{p}$ as

$$\bar{p} = (X'KK'X)^{-1}X'KK'y \qquad (A.1.29)$$

$$= (X'\Sigma^+X)^{-1}X'\Sigma^+y$$

if $(X'\Sigma^+X)^{-1}$ exists. The variance of $\bar{p}$ is

$$V(\bar{p}) = (X'KK'X')^{-1} = (X'\Sigma^+X)^{-1}. \qquad (A.1.30)$$

---

[4] Since we have transformed, T observations into g observations, there are $T - g$ observations implicitly discarded. These discarded observations are those associated with the $(T - g)$ zero value characteristic roots and the corresponding characteristic vectors. Let U_2 be the $(T \times (T - g))$ matrix which transforms the relation

$$U_2'y = U_2'Xp + U_2'u,$$

such that

$$EU_2'u = 0,$$

and

$$EU_2'uu'U_2 = 0.$$

As was explained in (A.1.18), the covariance of the disturbance of this transformation is zero since Σ is singular. The zero variance of $U_2'u$ and $EU_2'u = 0$ imply that $U_2'u = 0$ and thus the exact relation

$$U_2'y = U_2'Xp$$

is a linear equality constraint on the parameters. Therefore, the estimator (A.1.29) will be efficient if $U_2'X = 0$ so that the constraint vanishes (Theil 1964). Otherwise the constraint must be considered. In our Markov regression $U_2' = (I\,I \cdots I)$ is a $(T \times rT)$ matrix with r identities in a row. The constraint $U_2'y = U_2'Xp$, which does not vanish, is equivalent to $\eta_r = Gp$ when both sides are premultiplied by $(X_1'X_1)^{-1}X_1'$. This constraint has already been considered in the restricted least squares case. Even in the unrestricted case, the estimator (A.1.29) automatically fulfills the constraint because the observations are proportions. Therefore, in our case, the estimator is efficient.

In the case when Σ is nonsingular, then K equals H and (A.1.7) and (A.1.12) are identical. Therefore, (A.1.12) or (A.1.29) is a generalization of generalized least squares.

A.2. The generalized inverse of the disturbance covariance matrix for the Markov probability model

Consistent with proportional data, the disturbance covariance matrix is[5]

$$
\Sigma = \begin{bmatrix}
\dfrac{q_1(t)(1 - q_1(t))}{N(t)} & \dfrac{-q_1(t)\,q_2(t)}{N(t)} & \cdots & \dfrac{-q_1(t)\,q_r(t)}{N(t)} \\[2ex]
\dfrac{-q_2(t)\,q_1(t)}{N(t)} & \dfrac{q_2(t)(1 - q_2(t))}{N(t)} & \cdots & \dfrac{-q_2(t)\,q_r(t)}{N(t)} \\[2ex]
\vdots & \vdots & \ddots & \vdots \\[1ex]
\dfrac{-q_r(t)\,q_1(t)}{N(t)} & \dfrac{-q_r(t)\,q_2(t)}{N(t)} & \cdots & \dfrac{q_r(t)(1 - q_r(t))}{N(t)}
\end{bmatrix} \otimes I_{t=1}^{T} ,
$$

$$
= V(t) \otimes I_{t=1}^{T}, \tag{A.2.1}
$$

where $V(t)$ denotes a matrix whose diagonal elements are $q_i(t)(1 - q_i(t))/N(t)$ and off-diagonal elements, $-q_i(t)\,q_j(t)/N(t)$. The notation $I_{t=1}^{T}$ is defined in ch. 6. To obtain Σ^+, we obtain $V(t)^+$ since the following relation holds:

$$
\Sigma^+ = V(t)^+ \otimes I_{t=1}^{T}. \tag{A.2.2}
$$

To find $V(t)^+$, Σ^+, we may use the procedure described in the previous section. That is, first find the characteristic vectors associated with their positive characteristic roots to obtain U_1 and λ (arranged in the same order) and then derive K by dividing the ith column of U_1 by the square

---

[5] This matrix is discussed in chs. 5, 6, 7, 8 and 9.

root of the ith characteristic root. The generalized inverse is then KK'.[6]

When the size of Σ is large, the analytical derivation of Σ^+ is not very straightforward and therefore, in the following we apply Boot's (1963) method. The procedure is summarized as follows:

If A is singular, A can be written temporarily as

$$A = \begin{bmatrix} D & E \\ F & G \end{bmatrix}, \tag{A.2.3}$$

where D^{-1} exists and D has the same rank as A. Thus, the submatrix $(F\ G)$ is a linear combination of $(D\ E)$, and $(E\ G)$ is also a linear

---

[6] For example, consider a (2×2) matrix:

$$\Sigma_2 = \begin{bmatrix} \dfrac{p_1 p_2}{N} & \dfrac{-p_1 p_2}{N} \\ \dfrac{-p_1 p_2}{N} & \dfrac{p_1 p_2}{N} \end{bmatrix},$$

where $p_1 + p_2 = 1$. The characteristic roots are $2p_1 p_2/N$ and 0. The associated characteristic vectors when normalized are:

$$U_1 = \sqrt{1/2} \begin{bmatrix} 1 \\ -1 \end{bmatrix}, \quad \text{for } \lambda_1 = 2p_1 p_2/N,$$

and

$$U_2 = \sqrt{-1/2} \begin{bmatrix} 1 \\ 1 \end{bmatrix}, \quad \text{for } \lambda_2 = 0.$$

Thus, we have K as

$$K = \frac{\sqrt{N}}{\sqrt{p_1 p_2}} \begin{bmatrix} 1/2 \\ -1/2 \end{bmatrix}.$$

The generalized inverse is then

$$\Sigma_2^+ = KK' = \frac{\sqrt{N}}{\sqrt{p_1 p_2}} \begin{bmatrix} 1/2 \\ -1/2 \end{bmatrix} (1/2 \quad -1/2) \left(\frac{\sqrt{N}}{\sqrt{p_1 p_2}} \right)$$

$$= \begin{bmatrix} N/4p_1 p_2 & -N/4p_1 p_2 \\ -N/4p_1 p_2 & N/4p_1 p_2 \end{bmatrix}.$$

combination of $(D\ F)$. Let these linear scalars be H and K. Then A can be written as

$$A = \begin{bmatrix} D & E \\ F & G \end{bmatrix} = \begin{bmatrix} D & E \\ HD & HE \end{bmatrix} = \begin{bmatrix} D & DK \\ HD & HDK \end{bmatrix}$$

$$= \begin{bmatrix} I \\ H \end{bmatrix} D\ [I\ K] = RDQ', \tag{A.2.4}$$

where $R' = (I\ H)$ and $Q' = (I\ K)$. The generalized inverse A^+ is then given by

$$A^+ = Q\ (Q'Q)^{-1}D^{-1}(R'R)^{-1}R'. \tag{A.2.5}$$

Applying these formulas to our problem, we have for R and Q,

$$Q = R = \begin{bmatrix} I_{r-1} \\ -1\ -1\ \cdots\ -1 \end{bmatrix} = \begin{bmatrix} I_{r-1} \\ -\eta'_{r-1} \end{bmatrix}, \tag{A.2.6}$$

where η_{r-1} denotes a column vector with $r-1$ elements unity, and I_{r-1} is an $(r-1 \times r-1)$ identity matrix. The matrices Q and R are coincidently the same because $H = K = -\eta_r$. Thus, we have

$$Q'Q = R'R = [I_{r-1}\ \ -\eta_r] \begin{bmatrix} I_{r-1} \\ -\eta'_r \end{bmatrix} = [I_{r-1} + \eta_r\eta'_r] = [I_{r-1} + E_{r-1}], \tag{A.2.7}$$

where E_{r-1} denotes a $(r-1 \times r-1)$ matrix with all elements unity. Hence, $Q'Q$ or $R'R$ is a $(r-1 \times r-1)$ matrix with diagonal elements of 2 and 1 elsewhere. The determinant of $Q'Q$ is

$$\det Q'Q = r. \tag{A.2.8}$$

The cofactors $(Q'Q)_{ii}$ and $(Q'Q)_{ij}$ $(i \neq j)$ are

$$(Q'Q)_{ii} = r - 1 \tag{A.2.9}$$

and

$$(Q'Q)_{ij} = -1 \tag{A.2.10}$$

respectively.

The inverse of $Q'Q$ or $R'R$ is then

$$
(Q'Q)^{-1} = \begin{bmatrix} \dfrac{r-1}{r} & -\dfrac{1}{r} & -\dfrac{1}{r} & \cdots & -\dfrac{1}{r} \\[2mm] -\dfrac{1}{r} & \dfrac{r-1}{r} & -\dfrac{1}{r} & \cdots & -\dfrac{1}{r} \\[2mm] \vdots & \vdots & & \ddots & \vdots \\[2mm] -\dfrac{1}{r} & -\dfrac{1}{r} & -\dfrac{1}{r} & \cdots & \dfrac{r-1}{r} \end{bmatrix} = \left[I_{r-1} - \dfrac{1}{r} E_{r-1} \right].
$$

(A.2.11)

The products of $Q(Q'Q)^{-1}$ and $(R'R)^{-1}R'$ are the same and both equal to

$$
Q(Q'Q)^{-1} = \begin{bmatrix} I_{r-1} \\ \cdots \\ -\eta'_{r-1} \end{bmatrix} \begin{bmatrix} I_{r-1} - \dfrac{1}{r} E_{r-1} \end{bmatrix}
$$

$$
= \begin{bmatrix} \left(I_{r-1} - \dfrac{1}{r} E_{r-1} \right) \\ \cdots\cdots\cdots\cdots \\ \left(-\eta'_{r-1} + \dfrac{1}{r} \eta'_{r-1} E_{r-1} \right) \end{bmatrix} = \begin{bmatrix} \left(I_{r-1} - \dfrac{1}{r} E_{r-1} \right) \\ \cdots\cdots\cdots\cdots \\ \left(-\dfrac{1}{r} \eta'_{r-1} \right) \end{bmatrix}.
$$

(A.2.12)

On the other hand, the matrix D has rank $r-1$ and its inverse is[7]

$$
D^{-1} = \begin{bmatrix} \dfrac{N(t)}{q_1(t)} + \dfrac{N(t)}{q_r(t)} & \dfrac{N(t)}{q_r(t)} & \cdots & \dfrac{N(t)}{q_r(t)} \\[3mm] \dfrac{N(t)}{q_r(t)} & \dfrac{N(t)}{q_2(t)} + \dfrac{N(t)}{q_r(t)} & \cdots & \dfrac{N(t)}{q_r(t)} \\[3mm] \vdots & \vdots & \ddots & \vdots \\[3mm] \dfrac{N(t)}{q_r(t)} & \dfrac{N(t)}{q_r(t)} & \cdots & \dfrac{N(t)}{q_{r-1}} + \dfrac{N(t)}{q_r(t)} \end{bmatrix}
$$

$$
= \left[\dfrac{N(t)}{q_r(t)} E_{r-1} + A \right],
$$

(A.2.13)

---

[7] The inverse of D: see ch. 6.

where A denotes an $(r - 1 \times r - 1)$ diagonal matrix with elements $N(t)/q_i(t)$, $(i = 1, 2, ..., r - 1)$. Temporarily, let q_i be $q_i(t)$ and N be $N(t)$. By the use of formula (A.2.5), the generalized inverse of V is then

$$V^+(t) = \begin{bmatrix} \left(I_{r-1} - \dfrac{1}{r} E_{r-1} \right) \\ \hline \left(-\dfrac{1}{r} \eta'_{r-1} \right) \end{bmatrix} \times$$

$$\times \left[\dfrac{N}{q_r} E_{r-1} + A \right] \left[\left(I_{r-1} - \dfrac{1}{r} E_{r-1} \right) \left(-\dfrac{1}{r} \eta_{r-1} \right) \right]. \quad (A.2.14)$$

After the matrix multiplication performed on (A.2.14), if $V^+(t)$ is partitioned into four groups

$$V^+ = \begin{bmatrix} V_1 & V_2 \\ V_3 & V_4 \end{bmatrix}, \quad (A.2.15)$$

where V_1 is an $(r - 1 \times r - 1)$ matrix, V_2 and V_3 are $(r - 1 \times 1)$ and $(1 \times r - 1)$ vectors respectively, and V_4 is a scalar, then

$$V_1 = (N/q_r) E_{r-1} + A - ((r - 1)/r) E_{r-1} - (1/r) \eta_{r-1} B \quad (A.2.16)$$

$$- ((r - 1)/r) (N/q_r) E_{r-1} - (1/r) B' \eta_{r-1}$$

$$+ ((r - 1)^2/r^2) (N/q_r) E_{r-1} + (1/r^2) \eta_{r-1} B E_{r-1},$$

$$V_2 = V'_3 = -((r - 1)/r) (N/q_r) \eta_{r-1} - (1/r) B' \quad (A.2.17)$$

$$+ ((r - 1)^2/r^2) (N/q_r) \eta_{r-1} + (1/r^2) \sum_{i=1}^{r-1} (N/q_i) \eta_{r-1},$$

and

$$V_4 = ((r - 1)^2/r^2) (N/q_r) \eta_{r-1} + (1/r^2) B \eta_{r-1}, \quad (A.2.18)$$

where A is defined right after eq. (A.2.13) and B is a row vector $[N/q_1, N/q_2, ..., N/q_{r-1}]$. For purposes of clarification, we examine the case when $r = 2$. For this case, the matrix V^+ is then

$$V^+(t) = \begin{bmatrix} \dfrac{N}{4q_1 q_2} & -\dfrac{N}{4q_1 q_2} \\ -\dfrac{N}{4q_1 q_2} & \dfrac{N}{4q_1 q_2} \end{bmatrix}. \quad (A.2.19)$$

For the case when $r = 3$,

$$V^+(t) =$$

$$
\begin{bmatrix}
\dfrac{4}{9}\dfrac{N}{q_1} + \dfrac{1}{9}\dfrac{N}{q_2} + \dfrac{1}{9}\dfrac{N}{q_3} & -\dfrac{2}{9}\dfrac{N}{q_1} - \dfrac{2}{9}\dfrac{N}{q_2} + \dfrac{1}{9}\dfrac{N}{q_3} & -\dfrac{2}{9}\dfrac{N}{q_1} + \dfrac{1}{9}\dfrac{N}{q_2} - \dfrac{2}{9}\dfrac{N}{q_3} \\[3mm]
-\dfrac{2}{9}\dfrac{N}{q_1} - \dfrac{2}{9}\dfrac{N}{q_2} + \dfrac{1}{9}\dfrac{N}{q_3} & \dfrac{1}{9}\dfrac{N}{q_1} + \dfrac{4}{9}\dfrac{N}{q_2} + \dfrac{1}{9}\dfrac{N}{q_3} & \dfrac{1}{9}\dfrac{N}{q_1} - \dfrac{2}{9}\dfrac{N}{q_2} - \dfrac{2}{9}\dfrac{N}{q_3} \\[3mm]
-\dfrac{2}{9}\dfrac{N}{q_1} + \dfrac{1}{9}\dfrac{N}{q_2} - \dfrac{2}{9}\dfrac{N}{q_3} & \dfrac{1}{9}\dfrac{N}{q_1} - \dfrac{2}{9}\dfrac{N}{q_2} - \dfrac{2}{9}\dfrac{N}{q_3} & \dfrac{1}{9}\dfrac{N}{q_1} + \dfrac{1}{9}\dfrac{N}{q_2} + \dfrac{4}{9}\dfrac{N}{q_3}
\end{bmatrix}
$$

$$\text{(A.2.20)}$$

A.3. The multicollinearity case

Having the generalized inverse of Σ, we may proceed to obtain the estimator (A.1.12). In order to utilize formula (A.1.12), we must determine whether the $(r^2 \times r^2)$ matrix $(X'\Sigma^+X)^{-1}$ exists. Let the matrix Σ^+ be partitioned into blocks of diagonal submatrices s_{ij}

$$
\Sigma^+ = \begin{bmatrix}
s_{11} & s_{12} & \cdots & s_{1r} \\
s_{21} & s_{22} & \cdots & s_{2r} \\
\vdots & \vdots & \ddots & \vdots \\
s_{r1} & s_{r2} & \cdots & s_{rr}
\end{bmatrix},
\tag{A.3.1}
$$

where each s_{ij} is of size $(T \times T)$. The cross product $(X'\Sigma^+X)$ is then

$$
X'\Sigma^+X = \begin{bmatrix}
X_1's_{11}X_1 & X_1's_{12}X_2 & \cdots & X_1's_{1r}X_r \\
X_2's_{21}X_2 & X_2's_{22}X_2 & \cdots & X_2's_{2r}X_r \\
\vdots & \vdots & \ddots & \vdots \\
X_r's_{r1}X_1 & X_r's_{r2}X_2 & \cdots & X_r's_{rr}X_r
\end{bmatrix}.
\tag{A.3.2}
$$

Since in (A.3.2), Σ is singular, Σ^+ is also singular. The row sums of Σ^+ are zero. Moreover, since $X_1 = X_2 = \cdots = X_r = x$, the row sums of (A.3.2) are also zero

$$
\sum_{j=1}^{r} x's_{ij}x = x' \sum_{j=1}^{r} s_{ij}x = 0.
\tag{A.3.3}
$$

Thus, we conclude that $(X'\Sigma^+X)$ is singular and $(X'\Sigma^+X)^{-1}$ does not exist and hence, p cannot be solved by (A.1.12). This is the so called case of 'multi-collinearity'.

A.4. Row sum condition and the reduced weight matrix

As it is shown, the generalized inverse method in solving the Markov probability model encounters the problem of multicollinearity. However, even this case is not entirely hopeless. As illustrated by Johnston (1963, p. 202), if one can obtain independent estimates of an appropriate number of parameters, then the sample data can be employed to obtain estimates of the remaining parameters. Since we know our parameter p has the relation

$$\sum_j^r p_j = \eta_r, \tag{A.4.1}$$

if $r - 1$ of p_j's are known, the last vector p_r is also known by (A.4.1). Thus, we may regard the last T rows of $(X'\Sigma^+X)$ as linearly dependent on the first $(r-1)T$ rows, and instead of writing the complete set of normal equations,

$$(X'\Sigma^+X)\bar{p} = (X'\Sigma^+y), \tag{A.4.2}$$

we may write the abbreviated set

$$
\begin{bmatrix}
x's_{11}x & x's_{12}x & \cdots & x's_{1r}x \\
x's_{21}x & x's_{22}x & \cdots & x's_{2r}x \\
\vdots & \vdots & \ddots & \vdots \\
x's_{r-1,1}x & x's_{r-1,2}x & \cdots & x's_{r-1,r}x
\end{bmatrix}
\begin{bmatrix}
\bar{p}_1 \\
\bar{p}_2 \\
\vdots \\
\bar{p}_r
\end{bmatrix}
=
\begin{bmatrix}
\sum_i x's_{1i}y_i \\
\sum_i x's_{2i}y_i \\
\vdots \\
\sum_i x's_{r-1,i}y_i
\end{bmatrix}.
\tag{A.4.3}
$$

By the use of the proportion relations (A.4.1) and

$$y_r = \eta_r - \sum_{i=1}^{r-1} y_i, \tag{A.4.4}$$

the last r columns of the abbreviated set $(X'\Sigma^+X)$ may be expressed by the other columns. For instance, the set of equations in the first r rows of (A.4.3) is

$$x's_{11}xp_1 + x's_{12}xp_2 + \cdots + x's_{1r}xp_r = \Sigma_i \, x's_{1i}y_i, \quad \text{(A.4.5)}$$

which is equivalent to

$$x's_{11}xp_1 + x's_{12}xp_2 + \cdots + x's_{1r}x \, (\eta_r - p_1 - p_2 \cdots -p_{r-1})$$

$$= \sum_{i=1}^{r-1} x's_{1i}y_i + x's_{1r} \, (\eta_T - y_1 - y_2 \cdots -y_{r-1}). \quad \text{(A.4.6)}$$

After rearranging (A.4.6) becomes

$$x' \, (s_{11} - s_{1r}) \, x + x' \, (s_{12} - s_{1r}) \, x \cdots + x' \, (s_{1,r-1} - s_{1r}) \, x$$

$$= \sum_{i=1}^{r-1} x' \, (s_{1i} - s_{1r}) \, y_i. \quad \text{(A.4.7)}$$

Obviously, (A.4.3) may be rewritten as

$$\begin{bmatrix} x' \, (s_{11} - s_{1r}) \, x & \cdots & x' \, (s_{1,r-1} - s_{1r}) \, x \\ x' \, (s_{21} - s_{2r}) \, x & \cdots & x' \, (s_{2,r-1} - s_{2r}) \, x \\ \vdots & \ddots & \vdots \\ x' \, (s_{r-1,1} - s_{r-1,r}) \, x & \cdots & x' \, (s_{2,r-1} - s_{2r}) \, x \end{bmatrix} \begin{bmatrix} \bar{p}_1 \\ \bar{p}_2 \\ \vdots \\ \bar{p}_{r-1} \end{bmatrix}$$

$$= \begin{bmatrix} \sum\limits_{i}^{r-1} x' \, (s_{1i} - s_{1r}) \, y_i \\ \vdots \\ \sum\limits_{i}^{r-1} x' \, (s_{r-1,i} - s_{r-1,r}) \, y_i \end{bmatrix}. \quad \text{(A.4.8)}$$

Thus, the weight matrix has been reduced to

$$\Sigma^* = \begin{bmatrix} s_{11} - s_{1r} & & s_{1,r-1} - s_{1r} \\ s_{21} - s_{2r} & & s_{2,r-1} - s_{2r} \\ \vdots & & \vdots \\ s_{r-1} - s_{r-1,r} & \cdots & s_{r-1,r-1} - s_{r-1,r} \end{bmatrix}, \quad \text{(A.4.9)}$$

where the matrix Σ^* has the size $(r-1)T$ by $(r-1)T$. The abbreviated reduced weight matrix V^{*}[8] may be expressed in terms of $q_i(t)$ and $N(t)$ or temporarily q_i and N as

$$V^* = [V_1 - V_2\eta'_{r-1}] = [(1/r) \, (N/q_r) \, E_{r-1} + (A - (1/r)\eta_{r-1}B)], \quad \text{(A.4.10)}$$

---

[8] The abbreviated weight matrix V^* is defined as $\Sigma^* = V^* \otimes I^T_{t=1}$.

where V_1, V_2, A and B have been defined in § A.2. For example, when $r = 2$, the reduced weight matrix for the 2×2 case is

$$V^* = (1/2)(N/q_2) + (N/q_1) - (1/2)(N/q_1) = N/2q_1q_2, \quad \text{(A.4.11)}$$

and when $r = 3$, the reduced weight matrix for the 3×3 case is

$$V^* = (1/3)(N/q_3)\begin{bmatrix} 1 & 1 \\ 1 & 1 \end{bmatrix} + \begin{bmatrix} \dfrac{N}{q_1} & \\ & \dfrac{N}{q_2} \end{bmatrix} - (1/3)\begin{bmatrix} \dfrac{N}{q_1} & \dfrac{N}{q_2} \\ \dfrac{N}{q_1} & \dfrac{N}{q_2} \end{bmatrix}$$

$$= \begin{bmatrix} \dfrac{2}{3}\dfrac{N}{q_1} + \dfrac{1}{3}\dfrac{N}{q_3} & -\dfrac{1}{3}\dfrac{N}{q_2} + \dfrac{1}{3}\dfrac{N}{q_3} \\ -\dfrac{1}{3}\dfrac{N}{q_1} + \dfrac{1}{3}\dfrac{N}{q_3} & \dfrac{2}{3}\dfrac{N}{q_2} + \dfrac{1}{3}\dfrac{N}{q_3} \end{bmatrix}. \quad \text{(A.4.12)}$$

Thus, the unrestricted generalized inverse estimator may be obtained as

$$\bar{p}_* = (X'_* \Sigma^* X_*)^{-1} X'_* \Sigma^* y_*, \quad \text{(A.4.13)}$$

where X_* and y_* denote the abbreviated set of X and y with the last submatrix of vector deleted. In other words, X_* has only $r - 1$ blocks of X on diagonal and y_* has vectors y_1 to y_{r-1}. The last vector of p, p_r, may be obtained as

$$\bar{p}_r = \eta_r - \sum_{j=1}^{r-1} \bar{p}_j. \quad \text{(A.4.14)}$$

If one wishes to use the reduced weight matrix Σ^* as an experiment similar to that of ch. 5, one may have the unrestricted estimator given by (A.4.13). The restricted estimator may be obtained by minimizing

$$(y_* - X_* p_*) \Sigma^* (y_* - X_* p_*) \quad \text{(A.4.15)}$$

subject to the constraints

$$\sum_{j=1}^{r} p_j = \eta_r \quad \text{(A.4.16)}$$

and

$$p_j \geq 0, \quad j = 1, 2, ..., r - 1. \quad \text{(A.4.17)}$$

The quadratic programming simplex tableau is similar to that of ch. 6. However, unlike the generalized least squares or the maximum likelihood estimator, the restricted estimator weighted by Σ^* is not unique. In other words, if different sub-vectors of p are dropped and replaced by the residuals of the relation (A.4.16), the solution will be different. Only in the case of the two by two transition matrix, where the reduced weight is a scalar, is the solution the same as that obtained by the generalized least squares or the maximum likelihood method.

As an example showing the non-unique restricted solution obtained by (A.4.15), Telser's example (1963) is used. The unrestricted and restricted estimators weighted by the reduced matrix Σ^* of the generalized inverse Σ^+ are listed in table A.1 for the three different alternative sub-vectors of p dropped.

From table A.1, it is interesting to note that the unrestricted estimators are all the same regardless of which p_j is dropped. However, the

TABLE A.1

*Restricted and unrestricted estimator weighted by the reduced matrix
from the generalized inverse of the singular disturbance covariance matrix
for the Telser example*

Vector p_j dropped	Unrestricted estimators			Restricted estimators		
p_3	$\begin{bmatrix} 0.5748 & 0.2975 & 0.1277 \\ -0.0782 & 0.9791 & 0.0991 \\ 0.6195 & -0.3363 & 0.7168 \end{bmatrix}$			$\begin{bmatrix} 0.6140 & 0.1145 & 0.2144 \\ 0 & 0.8963 & 0.1037 \\ 0.4794 & 0 & 0.5206 \end{bmatrix}$		
p_2	$\begin{bmatrix} 0.5748 & 0.2975 & 0.1277 \\ -0.0782 & 0.9791 & 0.0991 \\ 0.6195 & -0.3363 & 0.7168 \end{bmatrix}$			$\begin{bmatrix} 0.6759 & 0.1320 & 0.1921 \\ 0 & 0.8783 & 0.1262 \\ 0.3979 & 0 & 0.6021 \end{bmatrix}$		
p_1	$\begin{bmatrix} 0.5748 & 0.2975 & 0.1277 \\ -0.0782 & 0.9791 & 0.0991 \\ 0.6195 & -0.3363 & 0.7168 \end{bmatrix}$			$\begin{bmatrix} 0.7692 & 0.1127 & 0.1181 \\ 0 & 0.8980 & 0.1020 \\ 0.2729 & 0 & 0.7271 \end{bmatrix}$		

restricted estimators are not unique since all specifications do not represent the proper condition for the minimization problem for all the parameters.

A.5. The unique solution of the generalized inverse estimator is the Aitken's generalized least squares with redundant parameters deleted

It is natural that now we investigate the non-unique solution of the restricted least squares weighted by Σ^*. Our investigation reveals that in order to obtain a unique solution the weight matrix should be symmetric no matter which set of parameters is dropped. The symmetric reduced weight matrix for this problem may be obtained as follows. Before the last set of normal equations (A.2.10) is dropped to write the abbreviated set as (A.2.11), we subtract the last set of normal equations from every set of normal equations to obtain the $r - 1$ sets of reduced normal equations as follows:

$$
\begin{bmatrix}
x' (s_{11} - s_{r1}) x & x' (s_{12} - s_{r2}) x & \cdots & x' (s_{1r} - s_{rr}) x \\
x' (s_{21} - s_{r1}) x & x' (s_{22} - s_{r2}) x & \cdots & x' (s_{2r} - s_{rr}) x \\
\vdots & \vdots & & \vdots \\
x' (s_{r-1,1} - s_{r1}) x & x' (s_{r-1,2} - s_{r2}) x & \cdots & x' (s_{r-1,r} - s_{rr}) x
\end{bmatrix}
\begin{bmatrix}
\bar{p}_1 \\
\bar{p}_2 \\
\vdots \\
\bar{p}_r
\end{bmatrix}
$$

$$
=
\begin{bmatrix}
\sum_i x' (s_{1i} - s_{ri}) y_i \\
\sum_i x' (s_{2i} - s_{ri}) y_i \\
\vdots \\
\sum_i x' (s_{r-1,i} - s_{ri}) y_i
\end{bmatrix} .
\tag{A.5.1}
$$

By doing so, we solve all equations simultaneously to insure the unique solution. Then, by the use of (A.4.1) and (A.4.4), we may express the last y_r by the other $r - 1$, y_i's and p_r by the other $r - 1$, p_i's. Eq. (A.5.1) becomes

$$
\begin{bmatrix}
x' b_{11} x & x' b_{12} x & \cdots & x' b_{1,r-1} x \\
x' b_{21} x & x' b_{22} x & \cdots & x' b_{2,r-1} x \\
\vdots & \vdots & & \vdots \\
x' b_{r-1,1} x & x' b_{r-1,2} x & \cdots & x' b_{r-1,r-1} x
\end{bmatrix}
\begin{bmatrix}
\bar{p}_1 \\
\bar{p}_2 \\
\vdots \\
\bar{p}_{r-1}
\end{bmatrix}
=
\begin{bmatrix}
\sum_i x' b_{1i} y_i \\
\sum_i x' b_{2i} y_i \\
\vdots \\
\sum_i x' b_{r-1,i} y_i
\end{bmatrix} ,
\tag{A.5.2}
$$

where

$$
b_{ij} = s_{ij} - s_{rj} - s_{ir} + s_{rr}.
\tag{A.5.3}
$$

That is, the weight matrix has been reduced to the well known symmetric matrix:

$$
\Sigma_*^{-1} = [b_{ij}] =
\begin{bmatrix}
\dfrac{N(t)}{q_r(t)} + \dfrac{N(t)}{q_1(t)} & \dfrac{N(t)}{q_r(t)} & \cdots & \dfrac{N(t)}{q_r(t)} \\[2ex]
\dfrac{N(t)}{q_r(t)} & \dfrac{N(t)}{q_r(t)} + \dfrac{N(t)}{q_2(t)} & \cdots & \dfrac{N(t)}{q_r(t)} \\[1ex]
\vdots & \vdots & & \vdots \\[1ex]
\dfrac{N(t)}{q_r(t)} & \dfrac{N(t)}{q_r(t)} & \cdots & \dfrac{N(t)}{q_r(t)} + \dfrac{N(t)}{q_{r-1}(t)}
\end{bmatrix}
$$

Thus, the set of normal equations becomes

$$(X'_* \Sigma_*^{-1} X_*)\, \bar{p}_* = X'_* \Sigma_*^{-1} y_* , \tag{A.5.4}$$

which is also the set of normal equations obtained from Aitken's generalized least squares on the model in which the last relation is deleted. The restricted estimator as shown in ch. 6 is unique.

A.6. Summary

As a summary to this appendix, we would like to clarify the relationship between the original model and the model in which a set of parameters is deleted. The following development enables us to explain how the matrix Σ_*^{-1} may be obtained from Σ^* and also from the diagonal weight matrix used in the minimum chi-square estimation.

First, let us consider the problem of maximizing

$$\phi(p) = -(y - Xp)'\, S\, (y - Xp), \tag{A.6.1}$$

subject to the constraints

$$Gp = \eta_r, \tag{A.6.2}$$

$$p \geq 0, \tag{A.6.3}$$

where y is a $(r \times 1)$ vector, X is a $(rT \times r^2)$ matrix with blocks of $(T \times r)$ submatrices x on the diagonal, p is a $(r^2 \times 1)$ vector with r subvectors p_j $(j = 1, 2, ..., r)$, each has size $(r \times 1)$, S is a weight matrix of size

$(rT \times rT)$, G is a $(r \times r^2)$ matrix with r identity matrix in a row and $\boldsymbol{\eta}_r$ is a $(r \times 1)$ vector with all units elements.[9]

The objective function (A.6.1) may be written in the form of the sum of the cross products

$$\phi(p) = -\sum_{i=1}^{r} (\boldsymbol{y}_i - x\boldsymbol{p}_i)' \, S_{ii} \, (\boldsymbol{y}_i - x\boldsymbol{p}_i)$$

$$-\sum_{i=1}^{r} \sum_{j \neq i}^{r} (\boldsymbol{y}_i - x\boldsymbol{p}_i)' \, S_{ij} \, (\boldsymbol{y}_j - x\boldsymbol{p}_j), \qquad \text{(A.6.4)}$$

where S_{ij} $(i, j = 1, 2, ..., r)$ are submatrices of S, each with size $(T \times T)$ and S may be partitioned as follows:

$$S = \begin{bmatrix} S_{11} & S_{12} & \cdots & S_{1r} \\ S_{12} & S_{22} & \cdots & S_{2r} \\ \vdots & \vdots & \ddots & \vdots \\ S_{r1} & S_{r2} & \cdots & S_{rr} \end{bmatrix}. \qquad \text{(A.6.5)}$$

The objective function (A.6.4) may also be written in two groups, the first $r - 1$ and the last r products

$$\phi(p) = -\sum_{i=1}^{r-1} (\boldsymbol{y}_i - x\boldsymbol{p}_i)' \, S_{ii} \, (\boldsymbol{y}_i - x\boldsymbol{p}_i) - (\boldsymbol{y}_r - x\boldsymbol{p}_r)' \, S_{rr} \, (\boldsymbol{y}_r - x\boldsymbol{p}_r)$$

$$-\sum_{i}^{r-1} \sum_{j \neq i}^{r-1} (\boldsymbol{y}_i - x\boldsymbol{p}_i)' \, S_{ij}(\boldsymbol{y}_j - x\boldsymbol{p}_j) - \sum_{j=1}^{r-1} (\boldsymbol{y}_r - x\boldsymbol{p}_r)' \, S_{rj} \, (\boldsymbol{y}_j - x\boldsymbol{p}_j)$$

$$-\sum_{i=1}^{r-1} (\boldsymbol{y}_i - x\boldsymbol{p}_i)' \, S_{ir} \, (\boldsymbol{y}_r - x\boldsymbol{p}_r). \qquad \text{(A.6.6)}$$

By the use of constraint (A.6.2), which is equivalent to

$$\sum_{j=1}^{r} p_j = \boldsymbol{\eta}_r, \qquad \text{(A.6.7)}$$

the objective function (A.6.6) may be reduced from r^2 parameters p to $(r - 1)r$ parameters p_*. Without loss of generality, let the last r parameters be replaced by the linear combination of the first $(r - 1)$ r para-

---

[9] These vectors and matrices are discussed in ch. 4.

meters. Thus we have

$$
\phi(p_*) = -\sum_{i=1}^{r-1} (y_i - xp_i)'\, S_{ii}\, (y_i - xp_i) - \sum_{i,j}^{r-1} (y_i - xp_i)'\, S_{rr}\, (y_j - xp_j)
$$

$$
-\sum_{i}^{r-1} \sum_{j \neq i}^{r-1} (y_i - xp_i)'\, S_{ij}(y_j - xp_j) + \sum_{i,j}^{r-1} (y_i - xp_i)'\, S_{rj}(y_j - xp_j)
$$

$$
+ \sum_{i,j}^{r-1} (y_i - xp_i)'\, S_{ir}\, (y_j - xp_j). \tag{A.6.8}
$$

In short, we may write (A.6.8) as

$$
\phi(p_*) = -\sum_{i=1}^{r-1} (y_i - xp_i)'\, (S_{ii} + S_{rr} - S_{rj} - S_{ir})\, (y_i - xp_i)
$$

$$
-\sum_{i}^{r-1} \sum_{j \neq i}^{r-1} (y_i - xp_i)'\, (S_{rr} + S_{ij} - S_{rj} - S_{ir})\, (y_j - xp_j)
$$

$$
= -\sum_{i,j}^{r-1} (y_i - xp_i)'\, (S_{rr} + S_{ij} - S_{rj} - S_{ir})\, (y_j - xp_j). \tag{A.6.9}
$$

Using compact matrix notation, we have

$$
\phi(p_*) = -(y_* - X_* p_*)'\, M\, (y_* - X_* p_*), \tag{A.6.10}
$$

where M is a $(r-1)\,T$ by $(r-1)\,T$ matrix in which the partitioned $(T \times r)$ submatrices M_{ij} are of the following form:

$$
M_{ij} = S_{rr} + S_{ij} - S_{rj} - S_{ir}. \tag{A.6.11}
$$

Thus, the diagonal blocks of matrices are $S_{rr} + S_{ii} - S_{ri} - S_{ir}$ and off diagonal blocks are $S_{rr} + S_{ij} - S_{rj} - S_{ir}$. This means that M may be obtained from S by using the following steps. Considering blocks of S_{ij} as elements, we may obtain M by subtracting the last column of S from every column of S and then subtracting the last row r of the resulted matrix from every row of the resulting matrix. Therefore, we have the following theorem:

To maximize $\phi(p) = -(y - Xp)'\, S\, (y - Xp)$ subject to the constraints $Gp = \eta_r$ and $p \geq 0$ is equivalent to maximizing $\phi(p_*)$ $= -(y_* - X_* p_*)'\, M\, (y_* - X_* p_*)$ subject to the constraints $Rp_* \leq \eta_r$ and $p_* \geq 0$, if and only if M is a matrix obtained from S such that $M_{ij} = S_{ij} + S_{rr} - S_{rj} - S_{ir}$, where M_{ij} and S_{ij} are T by T submatrices of M and S respectively, and R is a subset of G with the last identity matrix deleted, X, y and X_*, y_* are the matrices of observed proportions in the multivariate regression form for the Markov process, and the asterisk denotes a subset of a matrix or vector.

Examples of S which will yield $M = \Sigma_*^{-1}$ are Σ^+ and

$$\begin{bmatrix} N(t)/q_1(t) & & & \\ & N(t)/q_2(t) & & \\ & & \ddots & \\ & & & N(t)/q_r(t) \end{bmatrix} \otimes I_{t=1}^T. \qquad \text{(A.6.12)}$$

The latter is the weight matrix used in the minimum chi-square estimator, and the construction of Σ_*^{-1} from this matrix is very straightforward. To see the construction of Σ_*^{-1} from Σ^*, we use the formula (A.2.5) and write its cross section matrix $V(t)^+$ as

$$V(t)^+ = Q (Q'Q)^{-1} V(t)_*^{-1} (Q'Q)^{-1} Q'. \qquad \text{(A.6.13)}$$

If we pre- and post-multiply Q' and Q, we obtain

$$Q'V(t)^+ Q = V(t)_*^{-1}, \qquad \text{(A.6.14)}$$

or

$$Q'V(t)^+ Q \otimes I_{t=1}^T = \Sigma_*^{-1}, \qquad \text{(A.6.15)}$$

which states the exact construction procedure.

APPENDIX B

THE GENERAL LINEAR PROBABILITY MODEL

There are many instances in economics, sociology, biology and numerous other areas in which models are formulated to explain the variation of proportions, *i.e.*, discrete random variables. For example, in economics interest has ranged from the consideration of problems such as explaining the variation of dichotomous random variables such as "buy, not buy" or "use, not use", to multi-random variables such as the proportion of consumers in each category or proportions of firms in each size category. Some of the statistical models that have been formulated and corresponding empirical analysis for the dichotomous random variable case are reported in Zellner and Lee (1965).

B.1. The model

Assume that the tth observation for the jth category or state proportion, $y_j(t)$, based on $n_j(t)$ cases, is given by

$$y_j(t) = q_j(t) + u_j(t), \qquad t = 1, 2, ..., T \quad \text{and} \quad i, j = 1, 2, ..., r, \quad \text{(B.1.1)}$$

where $q_j(t)$ is the tth true proportion for the ith category and the $u_j(t)$ is assumed independently distributed, each with a multinomial distribution which has mean vector zero and variances $q_j(t) [1 - q_j(t)]/N(t)$ and covariances $-q_i(t) q_j(t)/N(t)$. The total number of cases is denoted by $N(t) = \Sigma_j n_j(t)$.

Assume now that the true proportions are related to determining or explanatory variables, at least over a range, by a relationship that is linear in the parameters such as

$$\boldsymbol{q}_i = X_i \boldsymbol{\gamma}_i, \qquad i = 1, 2, ..., r, \qquad \text{(B.1.2)}$$

where $\boldsymbol{q}_i$ is a $(T \times 1)$ vector of true proportions for the i category, X_i is a $(T \times K_i)$ matrix, with rank K_i, of observations on K_i nonstochastic variables, and γ_i is a $(K_i \times 1)$ unknown coefficient vector. Given eq. (B.1.1), eq. (B.1.2) may be rewritten in terms of the observed proportions as

$$y_i = X_i\gamma_i + u_i, \qquad i = 1, 2, ..., r, \qquad \text{(B.1.3)}$$

where y_i and u_i are now $(T \times 1)$ vectors with the properties defined above. The r equations in (B.1.3) may be written as the following set of equations:

$$\begin{bmatrix} y_1 \\ y_2 \\ \vdots \\ y_r \end{bmatrix} = \begin{bmatrix} X_1 & & & \\ & X_2 & & \\ & & \ddots & \\ & & & X_r \end{bmatrix} \begin{bmatrix} \gamma_1 \\ \gamma_2 \\ \vdots \\ \gamma_r \end{bmatrix} + \begin{bmatrix} u_1 \\ u_2 \\ \vdots \\ u_r \end{bmatrix} \qquad \text{(B.1.4)}$$

or compactly as

$$y = X\gamma + u. \qquad \text{(B.1.5)}$$

Since we are assuming the y_i are unbiased estimates of the q_i, i.e.,

$$E(y_i) = X_i\gamma_i = q_i, \qquad \text{for all } i, \qquad \text{(B.1.6)}$$

the u's have the properties that

$$E(u) = E(y - X\gamma) = q - X\gamma = 0, \qquad \text{(B.1.7)}$$

and

$$E(uu') = V(u) = V(y - X\gamma) = V(y). \qquad \text{(B.1.8)}$$

The variance–covariance of y for a particular observation period t is

$$V(t) = \frac{1}{N(t)} \begin{bmatrix} q_1(t)[1 - q_1(t)] & -q_1(t)q_2(t) & \cdots & -q_1(t)q_r(t) \\ -q_2(t)q_1(t) & q_2(t)[1 - q_2(t)] & \cdots & -q_2(t)q_r(t) \\ \vdots & \vdots & & \vdots \\ -q_r(t)q_1(t) & -q_r(t)q_2(t) & \cdots & q_r(t)[1 - q_r(t)] \end{bmatrix}. \qquad \text{(B.1.9)}$$

Because

$$q_1(t) + q_2(t) + \cdots + q_r(t) = 1, \qquad \text{(B.1.10)}$$

when analyzing data from a multinomial population, the number in any one cell may be disregarded because it is the sample size minus the total number in the other categories. Thus, we are left with $(r - 1)$ independent multinomial populations with mean vector q_* and dispersion matrix $V_*(t)$, a $((r-1) \times (r-1))$ submatrix of $V(t)$, (B.1.9) which, assuming $V_*(t)$ is full rank, has as its inverse,

$$
V_*(t)^{-1} = N(t)
\begin{bmatrix}
\dfrac{1}{q_1(t)} + \dfrac{1}{q_r(t)} & \dfrac{1}{q_r(t)} & \cdots & \dfrac{1}{q_r(t)} \\[2ex]
\dfrac{1}{q_r(t)} & \dfrac{1}{q_2(t)} + \dfrac{1}{q_r(t)} & \cdots & \dfrac{1}{q_r(t)} \\[1ex]
\vdots & \vdots & & \vdots \\[1ex]
\dfrac{1}{q_r(t)} & \dfrac{1}{q_r(t)} & \cdots & \dfrac{1}{q_{r-1}(t)} + \dfrac{1}{q_r(t)}
\end{bmatrix},
$$

$$(B.1.11)$$

where the asterisk denotes the $(r-1)$ independent categories. Thus, by considering the T observation periods the covariance matrix[1] for u_* is

$$E(u_* u'_*) = V_*(t) \otimes I, \tag{B.1.12}$$

if the sampling is from the Poisson scheme or

$$E(u_* u'_*) = V_*(t) \otimes I^T_{t=1}, \tag{B.1.13}$$

if the sampling is from the Lexis scheme.

B.2. The unrestricted estimator

If we let

$$E(u_* u'_*) = \Sigma_*, \tag{B.2.1}$$

and assuming Σ_* is of full rank, then Aitken's generalized least squares estimator may be applied to the model (B.1.4), (B.1.7) and (B.1.8). Under this specification, the best linear unbiased estimator of γ_*, the subset of γ, is

$$g_* = (X'_* \Sigma_*^{-1} X_*)^{-1} X'_* \Sigma_*^{-1} y_*, \tag{B.2.2}$$

when Σ_* is known. In practice, Σ_* is unknown and the usual way to proceed (Zellner and Lee 1965) is to use $\hat{\Sigma}_*$, a consistent estimate of the covariance matrix of y. Assuming that $\hat{\Sigma}_*$ is of full rank, the estimate

---

[1] See ch. 6 for the basis for the definition of the notation for these specifications.

$\hat{\Sigma}_*^{-1}$ may be obtained by replacing $y_i(t)$ for $q_i(t)$ in (B.1.11) and extended within the framework of (B.1.12) or (B.1.13). It should be noted that

$$g_* = (X_*'\hat{\Sigma}_*^{-1}X_*)^{-1}X_*'\hat{\Sigma}_*^{-1}y_* \tag{B.2.3}$$

is an approximation to the best linear unbiased estimator. The estimator (B.2.3) may be improved by a recursive approximation, in which the $q_i(t)$ of (B.1.11) will be replaced by $y_i(t)$ to obtain a new g_* and the next $\hat{y}_i(t)$ is obtained as

$$\hat{y}(t) = X_*g_*. \tag{B.2.4}$$

Assuming that S_* is of full rank at each stage, the iterative procedure may be repeated until two successive g_*'s are equal to an acceptable approximation. Unfortunately, the finite sample properties of the estimator are unknown.

The variance–covariance of the estimator g_* is

$$(X_*'\Sigma_*^{-1}X_*)^{-1}, \tag{B.2.5}$$

and may be estimated consistently by

$$(X_*'\hat{\Sigma}_*^{-1}X_*)^{-1}. \tag{B.2.6}$$

B.3. The restricted estimator

Since the predicted value of y_* is given by (B.2.4) and no restrictions have been imposed on either X_* or g_*, it is possible that $\hat{y}_*$ may fall outside interval $(0, 1)$. In order to avoid elements of $\hat{y}_*$ which are either negative or larger than unity, some restrictions may be necessary.

For the non-negativity of y_*, the appropriate constraints are

$$X_*g_* = \hat{y}_* \geq 0. \tag{B.3.1}$$

If X_* is known to be non-negative, then one way to insure that (B.3.1) is fulfilled is to impose non-negative constraints on γ_*. Alternatively, if X_* is known to be in the range (a, b), then possible combinations of a and b may be used to construct the restrictions.

The equal to or less than 1 constraint on $\hat{y}_*$ is the counterpart of (B.1.10) and may be reflected by the condition

$$G\hat{y}_*^c \leq \eta_T \tag{B.3.2}$$

or equivalently

$$GX_*g_*^c \leq \eta_T, \tag{B.3.3}$$

where G is a $(T \times (r-1)T)$ matrix $[I_1 I_2 \cdots I_{r-1}]$ with each I a $(T \times T)$ identity matrix and η_T is a column vector of dimension T with all elements equal to unity.

In order to eliminate unnecessary constraints, the sign of γ_* is not restricted. In formulating the restricted estimation problem within the framework of mathematical programming, the unknown coefficients γ_* are specified such that

$$\gamma_* = [\gamma_{*1} \vdots -\gamma_{*2}], \tag{B.3.4}$$

$$\gamma'_{*1}\gamma_{*2} = 0, \tag{B.3.5}$$

and

$$\gamma_{*1}, \gamma_{*2} \geq 0. \tag{B.3.6}$$

Thus, the original regression equation for the $(r-1)$ independent categories becomes

$$y_* = X_*\gamma_{*1} - X_*\gamma_{*2} + u_* \tag{B.3.7}$$

or compactly

$$y_* = Z_*\gamma_* + u_*, \tag{B.3.8}$$

with

$$Z_* = [X_* \vdots -X_*] \tag{B.3.9}$$

and

$$\gamma_* = \begin{pmatrix} \gamma_{*1} \\ \gamma_{*2} \end{pmatrix}. \tag{B.3.10}$$

Under this formulation, the problem is to minimize

$$\phi = u'_* u_* = (y_* - Z_*\gamma_*)' \Sigma_*^{-1} (y_* - Z_*\gamma_*) \tag{B.3.11}$$

subject to

$$\begin{cases} GX_*\gamma_{*1} - GX_*\gamma_{*2} \leq \eta_T, & \text{(B.3.12)} \\ \gamma_{*1}, \gamma_{*2} \geq 0, & \text{(B.3.13)} \end{cases}$$

or

$$\begin{cases} GZ_*\gamma_* \leq \eta_T, & \text{(B.3.14)} \\ \gamma_* \geq 0, & \text{(B.3.15)} \end{cases}$$

and (B.3.1) or

$$Z_*\gamma_* \geq 0. \tag{B.3.16}$$

Proceeding as in ch. 3, and subsequent chapters by making use of the reducibility theorem of non-linear programming, the problem becomes one of maximizing

$$(Z'_*\Sigma_*^{-1}y_* - Z'_*\Sigma_*^{-1}Z_*g_*^c)'\,\gamma_* \tag{B.3.17}$$

subject to

$$\begin{pmatrix} GZ_* \\ -Z_* \end{pmatrix}\gamma_* \leq \begin{pmatrix} \eta_T \\ 0 \end{pmatrix} \tag{B.3.18}$$

and

$$\gamma_* \geq 0. \tag{B.3.19}$$

Its dual problem is to minimize

$$\lambda' \begin{pmatrix} \eta_T \\ 0 \end{pmatrix} \tag{B.3.20}$$

subject to

$$\begin{pmatrix} GZ_* \\ -Z_* \end{pmatrix}'\lambda \geq Z'_*\Sigma_*^{-1}y_* - Z'_*\Sigma_*^{-1}Z_*\gamma_*, \tag{B.3.21}$$

and

$$\lambda \geq 0. \tag{B.3.22}$$

Within this context, the primal–dual formulation becomes one of maximizing

$$(Z'_*\Sigma_*^{-1}y_* - Z'_*\Sigma_*^{-1}Z_*\gamma_*)'\,\gamma_* - \lambda'\begin{pmatrix} \eta_T \\ 0 \end{pmatrix} \tag{B.3.23}$$

subject to

$$\begin{pmatrix} GZ_* \\ -Z_* \end{pmatrix}\gamma_* + w = \begin{pmatrix} \eta_T \\ 0 \end{pmatrix}, \tag{B.3.24}$$

$$\begin{pmatrix} GZ_* \\ -Z_* \end{pmatrix}'\lambda + Z'_*\Sigma_*^{-1}Z_*\gamma_* - v = Z'_*\Sigma'_*y_* \tag{B.3.25}$$

and

$$\lambda, \gamma_* \geq 0. \tag{B.3.26}$$

Rewriting (B.3.23) by the use of (B.3.24) and (B.3.25), we obtain

$$-\lambda^{\tau'}w^\tau - g_*^{c'}v^\tau = 0, \tag{B.3.27}$$

where the superscripts denote the optimal values. Eq. (B.3.27) simply means that since λ and w and γ_* and v are counterparts, then if the solution is optimal, at least one element of each pair must be zero.

Given this characteristic, the problem may be solved by the simplex procedure which has the tableau presented in table B.1.

If the tableau is expressed in terms of the original X_* and γ_*, etc. the following tableau results (table B.2).

TABLE B.1

The simplex tableau for the general linear probability model

p_0	λ_1	λ_2	γ_*	w_1	w_2	v
η_T			GZ_*	I_T		
0			$-Z_*$		I_T	
$Z_*'\Sigma_*^{-1}y_*$	$Z_*'G'$	$-Z_*'$	$Z_*'\Sigma_*^{-1}Z_*$			$-I_{2(r-1)}$

TABLE B.2

The simplex tableau for the general linear probability model

p_0	λ_1	λ_2	γ_{*1}	γ_{*2}	w_1	w_2	v_1	v_2
η_T			GX_*	$-GX_*$	I_T			
0			$-X_*$	X_*		I_T		
$X_*'\Sigma_*^{-1}y_*$	$X_*'G'$	$-X_*'$	$X_*'\Sigma_*^{-1}X_*$	$-X_*'\Sigma_*^{-1}X_*$			$-I_{r-1}$	
$-X_*'\Sigma_*^{-1}y_*$	$-X_*'G'$	X_*'	$-X_*'\Sigma_*^{-1}X_*$	$X_*'\Sigma_*^{-1}X_*$				$-I_{r-1}$

The last two constraints are a two way inequality, which means the equality should hold. For this case, the primal variable γ_* is not restricted in sign and the dual constraint results in the equality form. Therefore, the tableau may be simplified since v_1 and v_2 will always take zero values. w_1 is in fact the $y_r(t)$'s, the proportions deleted. Thus, the simplified tableau is as shown in table B.3.

TABLE B.3

The simplified simplex tableau for the general linear-probability model

p_0	λ_1	λ_2	γ_{*1}	γ_{*2}	y_{rt}	w_2
η_T			GX_*	$-GX_*$	I_T	
0			$-X_*$	X_*		I_T
$X_*'\Sigma_*^{-1}y_*$	$X_*'G'$	$-X_*'$	$X_*'\Sigma_*^{-1}X_*$	$-X_*'\Sigma_*^{-1}X_*$		

In table B.3, (λ_1, y_{rt}), (λ_2, w_2) and $(\gamma_{*1}, \gamma_{*2})$ are counterparts. The size of the tableau is $2T + \Sigma_{i=1}^{r-1} K_i$ by $4T + 2\Sigma_{i=1}^{r-1} K_i$.

Proceeding as before, if Σ_* is unknown, then it is replaced by a consistent estimate of the covariance matrix, $\hat{\Sigma}_*$, which is assumed to be of full rank, and the analysis goes forward as outlined in order to obtain estimates which fulfil restrictions (B.3.1) and (B.3.2).

B.4. A joint estimation procedure

As noted by Zellner and Lee (1965), the procedures discussed in §§ B.1, B.2 and B.3 can be applied to the problem of estimating a system of relationships each of which involve discrete random variables. Thus, we may have several sets of relations of the type specified in § B.1. If the discrete random dependent variables appearing in the different sets of equations are correlated with each other, then as noted by Zellner (1962), joint estimation procedures are asymptotically more efficient than co-efficient estimators that were proposed in the earlier sections. For the unrestricted estimators under this specification, the reader is referred to Zellner and Lee (1965). To insure that the estimated proportions over the sets of relations satisfy (B.3.1) and (B.3.2) for each set, a generalization of the formulation given in § B.3 is recommended.

ESTIMATION OF VARIABLE TRANSITION PROBABILITIES

The statistical models specified in this book are directed toward the problem of estimating the transition probabilities for a stationary Markov chain. Thus, the transition probabilities are assumed to be constant over the entire sample period. There are many cases, however, for which it is more plausible to assume that the transition probabilities are not invariant with time or over the sample period. Thus, it may be reasonable to assume that the transition probabilities are functions of certain explanatory variables and that they change as these variables change. For example, in investigating the cigarette consumption problem, Telser (1963) assumed that the probability of transition from one cigarette brand to another depended on the relative prices of the brands and on the relative advertising expenditures of the companies. Therefore, to handle this type of specification, we will discuss one way to estimate transition probabilities that are not constant over time when the data available for estimation purposes are time ordered observed proportions plus the observed values of some relevant explanatory variables.

C.1. The model

To generalize the stationary transition probability problem, let us write, using notation defined in earlier chapters, the first order non-stationary Markov chain statistical model as

$$y_j(t) = \sum_{i=1}^{r} y_i(t-1)\, p_{ij}(t) + u_j(t), \qquad j = 1, 2, ..., r, \quad (C.1.1)$$

where the parameters $p_{ij}(t)$ may change with time. If no information existed other than the first order non-stationary Markov chain specification (C.1.1), then it is not possible to estimate the variable parameters, $p_{ij}(t)$'s, because there are r^2 unknowns in every observation period. Previously, we assumed stationarity which means that

$$p_{ij}(t) = p_{ij}, \qquad \text{for all } t, i, j. \qquad (C.1.2)$$

Thus, the number of unknowns is reduced from r^2T to only r^2. As developed in ch. 3, these unknowns may then be estimated from the total of rT equations provided that T is at least equal to or greater than r.

For the variable transition probabilities, if we assume a linear relationship between the $p_{ij}(t)$ and some externally generated explanatory variables, $z_k(t)$, then we have additional information in the form of

$$p_{ij}(t) = \sum_{k=1}^{M} \delta_{ijk} z_k(t) + v_{ij}(t), \qquad \text{for all } i, j, t, \qquad (C.1.3)$$

where δ_{ijk} are parameters and $v_{ij}(t)$ is the stochastic element and $z_k(t)$ are the external explanatory variables. Thus, eq. (C.1.3) is a generalization of relation (C.1.2) where systematic and stochastic variation has been added to the previously time constant transition probabilities. Thus, the number of unknowns is reduced from r^2T to r^2M and they may be estimated from the rT observational equations provided that T is at least equal to or larger than rM. The introduction of explanatory variables in describing the variable transition probabilities has reduced the degrees of freedom. However, this reduction in the degrees of freedom may be compensated for by increasing the number of observations. As an aside, it should be noted that in applications of the stationary probability case, increasing the number of observations, (t), may increase the probability of rejecting the hypothesis of stationarity.

Since there are rT equations of type (C.1.1) and r^2T equations of type (C.1.3), we may write them in compact matrix notation as

$$y = Xp + u \qquad (C.1.4)$$

and

$$p = Z\delta + v, \qquad (C.1.5)$$

where y is a $(rT \times 1)$ vector of proportions $y_j(t)$, $j = 1, 2, ..., r$ and $t = 1, 2, ..., T$, u and v are $(rT \times 1)$ and $(r^2T \times 1)$ vectors of disturbances,

X is a $(rT \times r^2T)$ block diagonal matrix of proportions $y_i(t-1)$:

$$
X = \begin{bmatrix}
[y(t-1) \otimes I_T] & & & \\
& [y(t-1) \otimes I_T] & & \\
& & \ddots & \\
& & & [y(t-1) \otimes I_T]
\end{bmatrix},
$$

where each block is a $T \times rT$ matrix with $y(t) = (y_1(t)\, y_2(t) \cdots y_r(t))$ for $t = 0, 1, \cdots T-1$, p is a $(r^2T \times 1)$ vector of variable transition probabilities $p_{ij}(t)$ arranged as

$$
p = \begin{bmatrix} p_1 \\ p_2 \\ \vdots \\ p_r \end{bmatrix} \quad \text{with} \quad p_j = \begin{bmatrix} p_{1j} \\ p_{2j} \\ \vdots \\ p_{rj} \end{bmatrix} \quad \text{and} \quad p_{ij} = \begin{bmatrix} p_{ij}(1) \\ p_{ij}(2) \\ \vdots \\ p_{ij}(T) \end{bmatrix},
$$

Z is a $(r^2T \times r^2M)$ block diagonal matrix

$$
Z = \begin{bmatrix}
Z_1 & & & \\
& Z_2 & & \\
& & \ddots & \\
& & & Z_{r^2}
\end{bmatrix},
$$

with $Z_1 = Z_2 = \cdots = Z_{r^2}$, where each Z_i is a $(T \times M)$ matrix and δ is a $(r^2M \times 1)$ vector of parameters arranged as

$$
\delta = \begin{bmatrix} \delta_1 \\ \delta_2 \\ \vdots \\ \delta_r \end{bmatrix} \quad \text{with} \quad \delta_j = \begin{bmatrix} \delta_{1j} \\ \delta_{2j} \\ \vdots \\ \delta_{rj} \end{bmatrix} \quad \text{and} \quad \delta_{ij} = \begin{bmatrix} \delta_{ij1} \\ \delta_{ij2} \\ \vdots \\ \delta_{ijm} \end{bmatrix}.
$$

The stochastic assumptions for u and v are

$$Eu = 0, \tag{C.1.6}$$

$$Euu' = \Sigma, \tag{C.1.7}$$

$$Ev = 0, \tag{C.1.8}$$

$$Evv' = \Omega, \tag{C.1.9}$$

and

$$Euv = 0, \tag{C.1.10}$$

where Σ is a $(rT \times rT)$ diagonal block matrix with elements

$$q_i(t)\,(1 - q_i(t))/N(t)$$

on the diagonal of the diagonal blocks and $-q_i(t)\,q_j(t)/N(t)$ on the diagonal of the off diagonal blocks, the q's are the true proportions or the expected values of the y's, Ω is a $(r^2T \times r^2T)$ block diagonal matrix with elements $p_{ij}(t)\,(1 - p_{ii}(t))/n_i(t-1)$ on the diagonal of the diagonal blocks and $-p_{ij}(t)\,p_{ik}(t)/n_i(t-1)$ on the diagonal of the off diagonal blocks, $n_i(t-1)$ is equal to $N(t)\,y_i(t-1)$. Both matrices Σ and Ω are singular. Thus, our problem is to find δ such that the variable transition probabilities p are determined by Z when the first order Markov chain (C.1.4) holds.

By the substitution of (C.1.5) into (C.1.4), we obtain

$$y = XZ\delta + w, \tag{C.1.11}$$

where

$$w = Xv + u \tag{C.1.12}$$

and

$$Ew = XEv + Eu = 0, \tag{C.1.13}$$

$$Eww' = \omega = (X\Omega X') + \Sigma. \tag{C.1.14}$$

The covariance matrix ω is singular since both Σ and $X\Omega X'$ have row sums which are zero. This characteristic of Σ was discussed in ch. 5. The result follows for $X'\Omega X$ since it is a block diagonal matrix with elements $p_{ij}(t)\,(1 - p_{ij}(t))\,y_i(t-1)/N(t)$ on the diagonal blocks and $-p_{ij}(t)\,p_{ik}(t)\,y_i(t-1)/N(t)$ on the off-diagonal blocks. Thus, the whole matrix $X\Omega X' + \Sigma$ has row sums which are all zero. Proceeding in the usual way in order to mitigate the impact of a singular disturbance covariance matrix and using asterisks to denote the subsets of the matrices which result when the relations involving the last proportions y_r are deleted, we may write our abbreviated model as

$$y_* = X_*Z_*\delta_* + w_*, \tag{C.1.15}$$

where

$$Ew_* = 0, \tag{C.1.16}$$

$$Ew_*w_* = \omega_*, \tag{C.1.17}$$

and ω_* is non-singular with elements which are functions of $q_j(t)$ and

$p_{ij}(t)$. Although the formula for the inverse ω_*^{-1} is not analytically available, the numerical inversion may be obtained if $q_j(t)$ is replaced by $y_j(t)$ and $p_{ij}(t)$ is replaced by a neighbourhood value as an approximation. Successive approximations may then be obtained as described in chs. 8 and 9 for the maximum likelihood and Bayesian estimators. Other difficulties relate to finding the neighbourhood values of $p_{ij}(t)$ and the assurance of recursive convergency. On these topics, further study is in order.

If we can relax our assumption about the determination of variable transition probabilities in such a way that the variation of p is non-stochastic, then the stochastic term v vanishes and the covariance matrix ω_* is manageable. Of course, in this case, ω_* is the case as Σ_* whose inverse Σ_*^{-1} is a block diagonal matrix with elements $q_j(t)/N(t) + q_r(t)/N(t)$ on the diagonal blocks and $q_r(t)/N(t)$ on the off-diagonal blocks. In practice, the $y_j(t)$'s may be substituted for the $q_j(t)$'s and the recursive approximation procedure discussed in chs. 6, 8 and 9 may be applied. In the following development, we will use ω_* instead of Σ_* since the former, in fact, includes the latter.

C.2. The unrestricted estimator

Without imposing any restrictions, the generalized least squares estimator $\tilde{\delta}_*$ is

$$\tilde{\delta}_* = (Z_*' X_*' \omega_*^{-1} X_* Z_*)^{-1} Z_*' X_*' \omega_*^{-1} y_*, \qquad \text{(C.2.1)}$$

provided that $X_* Z_*$ has full column rank (i.e., the necessary but not sufficient condition is that T is larger than or equal to rM). The vectors of transition probabilities are then determined as

$$\hat{p}_*^e = Z_* \tilde{\delta}_* \qquad \text{(C.2.2)}$$

and

$$\hat{p}_r^e = \eta_{rT} - R\hat{p}_*^e, \qquad \text{(C.2.3)}$$

where η_{rT} is a $(rT \times 1)$ vector with unit elements and R is a $(rT \times r(r-1)T)$ matrix with $(r-1)$ identity matrices of size $(rT \times rT)$.

C.3. The restricted estimator

Since no restraint is imposed in (C.2.1), there is no guarantee that the transition probability p_* and p_r will be in the range between 0 and 1. Therefore, the restricted least squares estimator should be considered and the necessary constraints are

$$Rp_* \leq \eta_{rT} \tag{C.3.1}$$

and

$$p_* \geq 0. \tag{C.3.2}$$

When dealing with the parameter δ_*, the above constraints will have to be transformed into the following:

$$RZ_*\delta_* \leq \eta_{rT} \tag{C.3.3}$$

and

$$Z_*\delta_* \geq 0, \tag{C.3.4}$$

where δ_* itself is not restricted in sign. Therefore, we have the following quadratic programming problem:

Find $\tilde{\delta}_*^c$ which maximizes

$$-(y_* - X_*Z_*\delta_*)' \omega_*^{-1}(y_* - X_*Z_*\delta_*) \tag{C.3.5}$$

subject to (C.3.3) and (C.3.4). Since δ_* is not restricted in sign, we will form counterparts δ_Δ and $\delta_{\Delta\Delta}$ such that

$$\delta_* = \delta_\Delta - \delta_{\Delta\Delta}, \tag{C.3.6}$$

where the elements of δ_Δ and $\delta_{\Delta\Delta}$ are all non-negative and have the property

$$\delta_\Delta'\delta_{\Delta\Delta} = 0. \tag{C.3.7}$$

Thus, the original set of equations become

$$y_* = X_*Z_*\delta_\Delta - X_*Z_*\delta_{\Delta\Delta} + w_* \tag{C.3.8}$$

and the standard quadratic programming problem is:

To minimize

$$(y_* - X_*Z_*\delta_\Delta + X_*Z_*\delta_{\Delta\Delta})' \omega_*^{-1}(y_* - X_*Z_*\delta_\Delta + X_*Z_*\delta_{\Delta\Delta}) \tag{C.3.9}$$

subject to

$$RZ_*\delta_\Delta - RZ_*\delta_{\Delta\Delta} \leq \eta_{rT}, \qquad (C.3.10)$$

$$Z_*\delta_\Delta - Z_*\delta_{\Delta\Delta} \geq 0 \qquad (C.3.11)$$

and

$$\delta_\Delta, \delta_{\Delta\Delta} \geq 0. \qquad (C.3.12)$$

Developing the primal–dual formulation as described in appendix B or in chs. 4 and 6, the simplex tableau given in table C.1 results. In the tableau, λ_1 and λ_2 are dual variables, α_1 and α_2 are primal slack variables; λ_1 and α_1, λ_2 and α_2, and δ_Δ and $\delta_{\Delta\Delta}$ are counterparts. Thus, only one element of each pair may be in the basis.

TABLE C.1

The simplex tableau for determining the variable transition probabilities

B_0	λ_1	λ_2	δ_Δ	$\delta_{\Delta\Delta}$	α_1	α_2
η_{rT}			RZ_*	$-RZ^*$	I	
0			$-Z_*$	Z^*		I
$Z_*'X_*'\omega_*^{-1}y_*$	$Z_*'R'$	$-Z_*'$	$Z_*'X_*'\omega_*^{-1}X_*Z_*$	$-Z_*'X_*'\omega_*^{-1}X_*Z_*$		

C.4. Concluding remarks

In the above formulation, the transition probabilities are determined by $Z_*\tilde{\delta}_*^c$ and the restrictions are effective if the explanatory variables, Z, continue in the range of past observations. However, if the future values of the explanatory variables are uncontrolled, then there is still no guarantee that the predicted p may be in the range between one and zero. Thus, if the future values of the explanatory variables are known to be $\bar{Z}_*$, then additional restrictions of the type (C.3.10)–(C.3.12) may be constructed with the future values of Z_*. With valid future transition probabilities, the future proportions may also be predicted via the first order Markov chain model.

THE FORTRAN PROGRAMMING OF CLASSICAL AND BAYESIAN TRANSITION PROBABILITY ESTIMATORS

This program will compute alternative estimators of the transition probabilities. The possible alternative restricted and unrestricted estimators include classical, weighted, generalized inverse (first stage biased), two stage, minimum absolute deviation by linear programming, maximum likelihood and Bayesian posterior estimators.

By using the appropriate control card, the program will compute any alternative estimator appearing in this book with the option of printing intermediate results and punching the final transition matrix on cards. Prediction and the chi-square goodness-of-fit test may also be performed. For each single problem, only a few cards of observed proportions (or aggregate data) need to be punched. A sequence of problems may be computed at one time by repeating the input procedure, and the summary of the problems including means, variances, standard deviations, root mean square error, etc. may also be computed in a single run. However, the number of states is limited to 6 and the weight matrix must not exceed the size of (96×96).

D.1. The standard procedure

The typical input for the computation is the maximum likelihood estimator. The deck will consist of two control cards and the data deck. The first control card is punched "SAMPLE" in columns 1 to 6 signifying that the following cards are the sample information. Columns 9 to 72 may be used for the title or identification of the problem. The second control card is the card of "control of options", which will be discussed later (see section D.8). Following the two control cards are the data cards punched according to the variable format appearing in the second control card (control of options). The sample data cards must be punched with an augmented matrix of proportions or number of micro units in each state and the total number of observations in the last column. If the number of observations in all periods is the same, the alternative input procedure is to punch this constant number in columns 10 to 15 of the card of "control of options".

D.2. Incorporation of the prior knowledge with sample observations

If prior knowledge is going to be incorporated to obtain the Bayesian estimator, another control card punched "PRIORI" in columns 1 to 6 must follow the sample data deck. After the "PRIORI" card, the next card must be punched in columns 1 to 6 with the number of priori information cards to follow, and in columns 7 to 12, the options for using alternative prior knowledge: punch "0" for multivariate beta, "9999" for the independent univariate beta and "8888" for the normal prior. If the multi-beta prior or the univariate prior is used, the parameters I, J, a_{ij} and $a_i - a_{ij}$ must be punched according to the format (2I6, 2F15.4), where I is the number of rows, J the number of columns in the transition matrix, a_{ij} and $a_i - a_{ij}$ are the parameters of the beta distribution. If the normal prior is used, a_{ij} will be the prior mean and $a_i - a_{ij}$ will be replaced by the prior variance.

D.3. Deleting a column

The ML and the Bayesian estimators require deletion of one column of the proportions and the corresponding relation. Column 22 of the card of control of options is designed for this purpose. If no column is to be deleted, punch "0" in this column. If the number punched is larger than or equal to r, the number of Markov states, the rth column will be dropped. (Note: if "0" is used, the ML estimator will not be computed unless the recursive procedure is in order as described in section D.5.) If any column may be deleted, column r is recommended for easy reading of the results. The solution is unique for the ML estimator no matter which column is deleted. (It is also true for the Bayesian estimator.)

D.4. Assignment of weight

The weight matrix M in the objective function $\phi = (y - Xp)' \, M(y - Xp)$ may be assigned by the user. Column 23 of the card of control of options provides for the choice of weight. Most of the weight matrices are generated by the computer. However, if the user wishes to assign his own weight, he may punch "3" in this column. If the user does so, he must have another control card punched "WEIGHT" in columns 1 to 6 following the sample data deck. Then, each non-zero element of the weight matrix M must be punched on a card with the information I, J and VALUE according to the format (2I3, F15.4), where I and J are the number of rows and columns of M respectively, and VALUE is the assigned weight.

D.5. Recursive solutions for the ML and Bayesian estimators

The exact numerical solutions for the ML and Bayesian estimators may be obtained by a feedback procedure of recursive quadratic programming. To do this, a control card punched "RECURSIVE", beginning in column 1, should follow the deck of sample data cards or the priori cards depending on whether the recursive solution is an ML or a Bayesian estimator. After the "RECURSIVE" card, the next card must be punched in columns 4 to 6 with the number of recursive iterations required, and in column 12 the significant tolerance limit required (the format is 2I6). The tolerance limit is defined as the sum over all previous iterations of the absolute difference between two successive solutions. The recursive iterations will be terminated when either requirement is reached. From column 24 to 34, they are the same as that of the card of "control of options" (see section 8). This permits the user to assign printing options since the user may want to skip some of the intermediate output.

It is interesting to note that the initial solution for the exact ML estimate may not be necessary from the first approximated ML solution. That is, we may use, for example, the classical estimate with no column deleted as an initial solution. The recursive solutions will converge to the same final exact estimate. However, it is recommended that the first approximated ML solution be used as the initial solution to reduce the number of recursive iterations. The user also should note that the last column of the transition matrix will be deleted automatically when the "RECURSIVE" card is read by the computer.

D.6. The use of the control "DITTO" and "DIT."

For the Bayesian estimation, if a sequence of problems is to be computed with the same prior knowledge, the prior information must be provided after the card "PRIORI" for the first problem and the second problem may use "DITTO" or "DIT." instead of the "PRIORI" plus several cards of prior parameters. If "DITTO" is used, the list of the prior parameters will not be printed, while "DIT." will print the list for the second problem and so on. This option is particularly designed for the sampling experiment presented in this book.

D.7. The use of the control "CLEAR" and "SUMMARY"

This option is also designed primarily for the sampling experiment. If a set of problems is computed at one time in a sequence and the user wishes to summarize the estimates in terms of means, variances, standard deviations and true parameters, he may do so by using the "CLEAR" and "SUMMARY" control cards punched from columns 1 to 6 and 1 to 7 respectively. The arrangement is as follows: the "CLEAR" card should be at the beginning of the set of problems to be summarized. The card of the control of options must be punched "1" in column 35 for each problem participating in the summary. The card "SUMMARY" must be placed after the sequence of problems. The true parameter cards must be provided following the "SUMMARY" card with a format of (9F8.4). The number of parameters should not exceed 36, i.e., four cards with format (9F8.4). The true values of the transition matrix are arranged column by column.

D.8. Option control card

This card plays an important role in the estimation procedures, predictions and hypothesis testing. Printing options are controlled, and the list of the control elements are as follows:

Columns	Variable	Description
1 to 3		Blank.
4 to 6	NT1	Number of time periods.
7 to 9	NS	Number of Markov states.
10 to 15	SS	Sample sizes when they are identical in all periods, i.e., $N(t) = N(t-1) = N$. If in each time period, the $N(t)$'s are different, these columns must be blank. Instead, in the data deck, the sample size for each period should be punched after the proportions. The variable format in columns 37 to 72 must be so indicated. It is important to note: if these columns are not blank when they should be, the number appearing in these columns will supersede each sample size in each period. The observed units may be used instead of the observed proportions. See also the option of KEY(1) in column 26.
16 to 21	TOL	Tolerance used in matrix inversion and simplex iterations. If left blank, 1.0E-5 will be used.
22	KDROP	Column to be dropped.
23	KW	Assignment to weights, $KW = 0, 1, \ldots, 9$. Punch "0" for classical least squares "1" for ML (or GLS, MCS) estimator "2" for weighted by $N(t)/W_j(t)$ "3" for users to assign any weight matrix "4" for weighted by mean proportions "5" for weighted by products of mean proportions "6" for weighted by the derived weight from generalized inverse method (biases) "7" for weighted by $N(t)/(w_j(t)(1 - w_j(t))$ "8" unused "9" for two-stage weighted by estimated disturbance variance
24	KV	Printing option of the dispersion matrix or $(X'MX)^{-1}$. Punch "0" to skip printing "1" to print the matrix at all stages "2" to print the matrix for the first stage (unrestricted) only

Columns	Variable	Description

"3" to print the matrix for the second stage (restricted) only

"4" to print the matrix when the restrictions are all fulfilled in the unrestricted estimator

25   KP   Option of prediction and hypothesis testing.

Punch "0" not to predict

"1" to predict

"2" to predict for the first stage

"3" to predict for the second stage

"4" to predict for the acceptable estimator

26   KEY(1)   Option of printing the input data.

Punch "0" to print the observed units if provided

"1" to print the observed proportions

27   KEY(2)   Option of printing the weight matrix.

Punch "0" not to print the weight matrix

"1" to print the weight matrix

"2" to print the weight matrix for the second stage (estimated covariance matrix) only

28   KEY(3)   Option of printing cross product matrices $X'MX$ and $X'My$.

Punch "0" not to print

"1" to print at all stages

"2" to print for the first stage only

"3" to print for the second stage only

29   KEY(4)   Option of printing simplex tableaux.

Punch "0" not to print any simplex tableau

"1" to print the first simplex tableau for LP/QP

"7" to print the last simplex tableau for LP/QP

"8" to print the tableau when there is trouble such as linear dependency, degeneracy, looping, etc. for QP.

"9" to print each simplex tableau for small problems (This is time consuming. Do not use for long problems.)

30   KEY(5)   Option for computing determinant, iteration steps, and the right hand side of the simplex tableau (B_0 column).

	Determinant	Iteration	B_0 Column
Punch "0"	no	no	no
"1"	no	no	yes
"2"	no	yes	yes
"3"	yes	no	no
"4"	yes	no	yes
"5"	yes	yes	yes

Columns	Variable	Description
31 to 32	KEY(6)	Scaling factor.
		Punch "1" for 1.0E01 and "2" for 1.0E02, etc.
33	KEY(7)	Option of computing the restricted estimator.

Punch "0" not to compute the restricted estimator

"1" to compute the restricted estimator for the first stage only

"2" to compute the restricted estimator for the second stage only

"3" to compute the restricted estimators for all stages but invalid for $KW = 9$

"4" to compute the restricted estimator for all stages but restriction of printing in second stage according to KEY(4)

"5" for all stages but restriction of printing in second stage according to KEY(4)

| 34 | KEY(8) | Choice of iteration procedure. |

Punch "0" to solve the problem by LP weighted by square root of column sum of M

"1" to solve the problem by LP weighted by column sum of M

"2" to solve the problem by QP

"9" to solve the problem by QP for the Bayesian estimator only

35	KEY(9)	Punch "1" to participate in the summary otherwise will be neglected
36		Blank.
37 to 72	FMT	Variable format of the data cards beginning with the left parenthesis and ending with the right parenthesis.

D.9. An input example

The numerical example used in chapters 7 and 9 will be used as an example of the input sequence. The following are the control and data cards:

```
SAMPLE           AN ABSORBING CHAIN
  005002   100      21041008200320    (4X, 2F8.4)
       0.5000    0.5000
       0.7500    0.2500
       0.8800    0.1200
       0.9400    0.0600
       0.9700    0.0300
```

RECURSIVE

6	3	041008200320

PRIOR I

4

1	1	99.0	1.0
1	2	1.0	99.0
2	1	50.0	50.0
2	2	50.0	50.0

RECURSIVE

6	3	041008200320

FINIS

D.10. An output example

If the above input example is used, the following output results:

AN ABSORBING CHAIN
THE OBSERVED PROPORTIONS AND SAMPLE SIZES

	1		2		3	
1	0.5000000E	00	0.5000000E	00	0.1000000E	03
2	0.7500000E	00	0.2500000E	00	0.1000000E	03
3	0.8800000E	00	0.1199999E	00	0.1000000E	03
4	0.9700000E	00	0.3000000E	00	0.1000000E	03

AS YOU DESIRED, COLUMN 2 IS DROPPED IN FORMIN G THE SYSTEM.
READ THE FOLLOWING MATRICES IN THE SEQUENCE 1 2

THE WEIGHT MATRIX IS DERIVED FROM THE ML E WITH ELEMENTS $N/WR(T)$ & $Z*(N/WJ(T))$, $Z = 1$ FOR DIAGONALS AND $Z = 0$ OTHER-WISE.

UNRESTRICTED ESTIMATOR OF THE TRANSITION MATRIX

0.9999978 5	0.00000215
0.505143 52	0.49485648

THE PREDICTED PROPORTIONS

	1		2	
2	0.7525707E	00	0.2474293E	00
3	0.8762842E	00	0.1237157E	00
4	0.9406152E	00	0.5938 463 E–01
5	0.9703066E	00	0.2969 340E–01
6	0.9851522E	00	0.1484 777E–01

SUM OF SQUARED ERROR	MEAN SQUARED ERROR	CHI SQUARE VALUE
0.000041776	0.000005222	0.017288841

MODIFIED CHI SQUARE
0.017593563

THE UNRESTRICTED ESTIMATOR IS PERFECT.

RECURSIVE OUTPUT, $K = 1$

UNRESTRICTED ESTIMATOR OF THE TRANSITION MATRIX
0.99999774   0.00000226
0.50501567   0.49498433

THE PREDICTED PROPORTIONS

	1	2
2	0.7525067E 00	0.2474933E 00
3	0.8762522E 00	0.1237477E 00
4	0.9405998E 00	0.5940008E–01
5	0.9702988E 00	0.2970118E–01
6	0.9851482E 00	0.1485172E–01

SUM OF SQUARED ERROR   MEAN SQUARED ERROR
0.000041557   0.000005195

CHI SQUARE VALUE   MODIFIED CHI SQUARE
0.017281160   0.017597124

THE UNRESTRICTED ESTIMATOR IS PERFECT.

RECURSIVE ERROR = 0.000256

BAYESIAN ESTIMATION OF THE TRANSITION PROBABILITIES

PRIOR PARAMETERS $A(I, J)$
99.000000   1.000000
50.000000   50.000000

MULTI-BETA PRIOR MEANS
0.990000   0.010000
0.500000   0.500000

MULTI-BETA PRIOR COVARIANCE MATRIX

	1	2
1	0.9801979E–04	0.0
2	0.0	0.2475247E–02

UNRESTRICTED ESTIMATOR OF THE TRANSITION MATRIX
1.00015163   −0.00015163
0.50168413   0.49831587

THE UNRESTRICTED ESTIMATOR VIOLATES THE PROPERTIES OF PROBABILITY. HENCE, THE ITERATION PROCEDURE IS REQUIRED.

THE FOLLOWINGS ARE QP SOLUTIONS

NO. OF ITERATIONS	PIVOT ROW	NEW BASIS	OLD BASIS	OBJ. VALUE
1	1	3	5	0.1635901E 05
2	4	4	0	0.6617109E 03
3	3	1	0	0.2175781E 01

NUMBER OF CYCLES REQUIRED = 3,
OBJ. VALUE FOR LP OR ROUNDING ERROR FOR QP = $-0.2685547E-02$

I	J	B(I)
1	3	1.0000000000
2	6	0.4981471896
3	1	2.1784667969
4	4	0.5018528104

HENCE, THE BAYESIAN ESTIMATOR OF THE TRANSITION MATRIX IS

1.00000000   0.0
0.50185281   0.49814719

RECURSIVE OUTPUT, $K = 1$
MULTI-BETA PRIOR COVARIANCE MATRIX

	1	2
1	0.9801979E-04	0.0
2	0.0	0.2475247E-02

UNRESTRICTED ESTIMATOR OF THE TRANSITION MATRIX

1.00000000   0.0
0.50184596   0.49815404

THE PREDICTED PROPORTIONS

	1	2
2	0.7509230E 00	0.2490770E 00
3	0.8754615E 00	0.1245385E 00
4	0.9402214E 00	0.2988924E-01
5	0.9701107E 00	0.2988924E-01
6	0.9850553E 00	0.1494462E-01

SUM OF SQUARED ERROR	MEAN SQUARED ERROR
0.000043023	0.000005378

CHI SQUARE VALUE	MODIFIED CHI SQUARE
0.019477647	0.020089485

THE UNRESTRICTED ESTIMATOR IS PERFECT.

RECURSIVE ERROR = 0.000014

D.11. Fortran listing of computer routine

This following computer routine permits the estimation of the Markov transition probabilities by the unweighted, weighted, generalized inverse, maximum likelihood, Bayesian and minimum absolute deviations estimators. In addition, the predicted proportions are generated and the chi-square goodness-of-fit test is computed.[1]

```
C
      DIMENSION A(30,30), B(30,2), C(30,30), D(30), E(30,30), X(96,30),
     1Y(96), S(96,96), XS(30,96), KEY(9), NP(96), FMT(9), IS(17), PP(16)
      COMMON A, B, C, D, E, X, Y, XS, S, SS, TOL, KDROP, KW, KV, KP,
     1KEY, LD, LR, LC, LX, MM, NN, NP, NS, NS1, NT, NT1, NOCYCL
C
      DATA SAMPLE/'SAMP'/
      DATA PRIORI/'PRIO'/
      DATA DITTO/'DITT'/
      DATA DITO/'DIT'/
      DATA CLEAR/'CLEA'/
      DATA SUMMAR/'SUMM'/
      DATA RECURS/'RECU'/
      DATA FINIS/'FINI'/
C     READ SAMPLE INFORMATION
  105 FORMAT (18A4)
  110 FORMAT (1H1)
  112 FORMAT (51H0THIS PROBLEM EXCEEDS THE CAPACITY OF
     1THIS PROGRAM./)
  114 FORMAT (3X, 2I3, F6.0, E6.0, 9I1, I2, 3I1, 1X, 9A4)
  115 READ (5,105) TESTA, (IS(I), I = 1, 17)
  116 IF (TESTA-SAMPLE)  117, 125, 117
  117 IF (TESTA-CLEAR)   120, 121, 120
  120 IF (TESTA-FINIS)   115,  99, 115
  121 NS = 6
      NS1 = 5
      CALL SMRY(1,1)
      GO TO 115
  125 READ (5,114) NT1, NS, SS, TO, KDROP, KW, KV, KP, (KEY(I), I = 1,9),
```

---

[1] Since the following listing is not a photo copy of the original and due to the fact that each line of this book has less than 80 characters, the character counts on the Hollerith field specifications are not precise. Thus, errors may be expected in format and continuation statements. However, the program logic is correct.

```
        1(FMT(I), I = 1,9)
          IF (TO) 126, 126, 127
    126 TOL = 1.0E-6
          GO TO 128
    127 TOL = TO
    128 WRITE (6,110)
          WRITE (6,105) (IS(I), I = 2,17)
          IF (KDROP) 129, 129, 130
    129 NS1 = NS
          GO TO 131
    130 NS1 = NS - 1
    131 NT = NT1 - 1
          MM = NS1*NT
          IF (NS-6) 132, 132, 133
    132 IF (MM-96) 134, 134, 133
    133 WRITE (6,112)
          GO TO 115
C       GENERALIZATION OF N INTO N(T)
    134 NS2 = NS + 1
          IF (SS) 80, 80, 135
     80 K = NS + 1
          GO TO 136
    135 K = NS
    136 DO 140 I = 1,NT1
          READ (5,FMT) (A(I,J), J = 1,K)
          IF (KEY(1) - 1) 140, 137, 137
    137 ROWSUM = 0.0
          DO 138 J = 1,NS
    138 ROWSUM = ROWSUM + A(I,J)
          DO 139 J = 1,NS
    139 A(I,J) = A(I,J)/ROWSUM
    140 CONTINUE
          IF (SS) 143, 143, 141
    141 DO 142 I = 1,NT1
    142 A(I,NS2) = SS
    143 IF (KEY(1) - 1) 145, 160, 160
    145 DO 146 I = 1,NT1
    146 A(I,NS2) = A(I,NS2)*A(I,NS2)
          WRITE (6,156)
    156 FORMAT (36H0THE OBSERVED UNITS AND SAMPLE SIZES/)
          GO TO 165
    160 WRITE (6,163)
    163 FORMAT (42H0THE OBSERVED PROPORTIONS AND SAMPLE
        1SIZES/)
```

```
      165 CALL PRINT (1, 1, NT1, NS2)
          IF (KEY(8) − 1) 166, 166, 170
C         SOLVED BY LINEAR PROGRAMMING
      166 LC = 2*NS1*NT + NS*NS
          IF (LC−96) 167, 167, 133
      167 CALL LP
          GO TO 115
      170 IF (KDROP) 180, 180, 190
C         TAKING CARE OF THE WEIGHT AND SOLVING FOR P.
      180 CALL OMEGA(2)
          GO TO 191
      190 CALL OMEGA(1)
      191 IF (KW−9) 500, 192, 500
      192 WRITE (6,193)
      193 FORMAT (48H0THE FOLLOWINGS ARE THE SECOND STAGE
          1ESTIMATION.)
          CALL PREDIC (10)
          DO 197 J = 1,NS1
          X(1,J) = 0.0
          DO 197 I = 2,NT1
      197 X(1,J) = X(1,J) + (A(I,J) − E(I,J))**2
          DO 198 I = 1,NT1
          DO 198 J = 1,NS
      198 A(I,J) = E(I,J)
          IF (KEY(7) − 2) 202, 200, 202
      200 WRITE (6,201)
      201 FORMAT (50H0THE DISTURBANCES ARE FROM UNRESTRICT-
          1ED ESTIMATOR.)
          GO TO 205
      202 WRITE (6,203)
      203 FORMAT (48H0THE DISTURBANCES ARE FROM RESTRICTED
          1ESTIMATOR.)
      205 KEY(7) = KEY(7) − 1
          KW = 10
          CALL OMEGA(3)
C         READ BETA PRIOR OR NORMAL PRIOR
      500 READ (5,105) TESTA, (IS(I), I = 1, 17)
          IF (TESTA-RECURS) 700, 703, 700
      700 IF (TESTA-PRIORI) 507, 511, 507
      507 IF (TESTA-DITTO) 508, 510, 508
      508 IF (TESTA-DITO) 600, 510, 600
      510 CALL BETA(TESTA,1)
          GO TO 701
      511 CALL BETA(TESTA,2)
```

```
  701  READ (5,105) TESTA, (IS(I), I = 1, 17)
       IF (TESTA-RECURS) 700, 702, 700
C         RECURSIVE ML AND BAYESIAN ESTIMATOR
  702  NBB = 1
       GO TO 704
  703  NBB = 0
  704  READ (5,799) NREC, NST, KV, KP, (KEY(I), I = 1, 9)
  799  FORMAT (5X, I1, 5X, I1, 11X, 7I1, I2, 3I1)
       IF (KDROP) 800, 800, 801
  800  NS1 = NS - 1
       MM = NS1*NT
       KDROP = NS
  801  KW = 1
       TT = 0.1**NST
       L = NS*NS
       DO 705 I = 1, L
  705  PP(I) = B(I,1)
       DO 752 KREC = 1, NREC
       WRITE (6,706) KREC
  706  FORMAT (21H1RECURSIVE OUTPUT, K = , I5/)
       CALL PREDIC(10)
       CALL SIGMA(7)
       IF (NBB-1) 707, 710, 707
  707  CALL OMEGA(4)
       GO TO 740
  710  CALL OMEGA(5)
C         AMENDING, STORE 0.00001 FOR P(I,J) = 0.0
       DO 733 I = 1, NS
       DO 733 J = 1, NS
       K = I + NS*(J - 1)
       IF (PP(K)) 732, 731, 732
  731  XS(I,J) = 0.00001
       GO TO 733
  732  XS(I,J) = PP(K)
  733  CONTINUE
       CALL BETA (TESTA, 3)
  740  RR = 0.0
       DO 748 I = 1, L
       RR = RR + ABS(PP(I) - B(I,1))
  748  PP(I) = B(I,1)
       WRITE (6,750) RR
  750  FORMAT (23H0RECURSIVE DIFFERENCE =, F9.6)
       IF (RR-TT) 500, 500, 752
  752  CONTINUE
```

```
      GO TO 500
  600 IF (TESTA-SUMMAR) 116, 610, 116
  610 CALL SMRY(3,1)
      CALL SMRY(3,2)
      GO TO 500
   99 CALL EXIT
      END
C
C
C     THIS SUBROUTINE INCORPORATES UNIVARIATE BETA, OR
C     MULTIVARIATE BETA, OR INDEPENDENT NORMAL WITH
C     SAMPLE OBSERVATION TO OBTAIN POSTERIOR ESTIMATES.
C
      SUBROUTINE BETA(TESTA,IB)
C
      DIMENSION A(30,30), B(30,2), C(30,30), D(30), E(30,30), X(96,30),
     1Y(96), S(96,96), XS(30,96), KEY(9), NP(96), PR(6,7), PQ(6,6)
      COMMON A, B, C, D, E, X, Y, XS, S, SS, TOL, KDROP, KW, KV, KP
     1KEY, LD, LR, LC, LX, MM, NN, NP, NS, NS1, NT, NT1, NOCYCL
C
      DATA DITTO/'DITT'/
      DATA DITO/'DIT.'/
      GO TO (510, 511, 834), IB
  510 IF (KALT-8888) 826, 901, 826
  511 WRITE (6,501)
      ARBI = 10.0**KEY(6)
      READ (5,503) NBETA, KALT
      IF(KALT-9999) 800, 522, 800
  800 IF(KALT-8888) 801, 512, 801
C     FROM UNIVARIATE BETA PRIOR TO MULTIVARIATE BETA
C     PRIOR PDF
  801 SN = NS
      DO 810 I = 1, NS
  810 PR(I,7) = 0.0
      DO 820 NPR = 1, NBETA
  820 READ (5,503) I, J, PR(I,J), PQ(I,J)
      DO 825 I = 1, NS
      DO 825 J = 1, NS
  825 PR(I,7) = PR(I,7) + PR(I,J)
  826 DO 828 I = 1, NS
      DO 828 J = 1, NS
  828 XS(I,J) = PR(I,J)/PR(I,7)
      IF (TESTA-DITTO) 831, 834, 831
  831 WRITE (6,870)
```

```
      DO 829 I = 1, NS
829   WRITE (6,873) (PR(I,J), J = 1, NS)
      WRITE (6,871)
      DO 830 I = 1, NS
830   WRITE (6,873) (XS(I,J), J = 1, NS)
834   L = NS*NS
      DO 832 I = 1, L
      DO 832 J = 1, L
832   A(I,J) = 0.0
      DO 850 I = 1, NS
      DO 850 J = 1, NS1
      K = I + NS*(J - 1)
835   C(K,K) = C(K,K) + (PR(I,7) - SN)/(XS(I,J)*ARBI) + (PR(I,7) - SN)/
     1(XS(I,NS)*ARBI)
      D(K) = D(K) + (PR(I,J) - 1.0)/(XS(I,J)*ARBI) + (PR(I,7) -
     1PR(I,NS) - SN + 1.0)/(XS(I,NS)*ARBI)
      DO 850 JK = J, NS
      IF (JK-J) 850, 850, 837
837   L = I + NS*(JK - 1)
      IF (JK - NS) 839, 845, 845
839   C(K,L) = C(K,L) + (PR(I,7) - SN)/(XS(I,NS)*ARBI)
840   C(L,K) = C(K,L)
845   IF (TESTA-DITTO) 846, 850, 846
846   A(K,K) = PR(I,J)*(PR(I,7) - PR(I,J))/(PR(I,7)**2*(PR(I,7) + 1.0))
      A(K,L) = -PR(I,J)*PR(I,JK)/(PR(I,7)**2*(PR(I,7) + 1.0))
      A(L,K) = A(K,L)
850   CONTINUE
      IF (TESTA-DITTO) 855, 526, 855
855   WRITE (6,872)
      CALL PRINT (1,1,K,K)
870   FORMAT (24H0PRIOR PARAMETERS A(I,J)/)
871   FORMAT (23H0MULTI-BETA PRIOR MEANS/)
872   FORMAT (35H0MULTI-BETA PRIOR COVARIANCE MATRIX/)
873   FORMAT (1H, 6F12.6)
      GO TO 526
901   WRITE (6,501)
      WRITE (6,509)
      DO 521 M = 1, NS
      DO 521 N = 1, NS1
      R = PR(M,N)
      Q = PQ(M,N)
      ARBI = 10.0**KEY(6)
      K = M + NS*(N - 1)
      GO TO 514
```

```
512 WRITE (6,502)
    DO 520 NPR = 1, NBETA
    READ 503, I, J, R, Q
    PR(I,J) = R
    PQ(I,J) = Q
    PM = R/(R + Q)
    V = (R*Q)/((R + Q + 1.0)*(R + Q)**2.0)
    IF (J - NS1) 513, 513, 520
513 K = I + NS*(J - 1)
514 IF (KEY(6)) 516, 515, 516
515 C(K,K) = C(K,K) + (R + Q)**2.0*(R + Q - 2.0)/(R*Q)
    D(K) = D(K) + (R + Q)**2.0*(R - 1.0)/(R*Q)
    GO TO 517
516 C(K,K) = C(K,K) + (R + Q)**2.0*(R + Q - 2.0)/(R*Q*ARBI)
    D(K) = D(K) + (R + Q)**2.0*(R - 1.0)/(R*Q*ARBI)
517 IF (TESTA-DITTO) 518, 521, 518
518 IF (TESTA-DITO) 520, 519, 520
519 I = M
    J = N
520 WRITE (6,503) I, J, R, Q, PM, V
521 CONTINUE
    GO TO 526
C   ASSUMING NORMAL PRIOR
522 WRITE (6,505)
    DO 525 NPR = 1, NBETA
    READ (5,506) I, J, AVE, VAR
    IF (J - NS1) 555, 555, 525
555 K = I + NS*(J - 1)
    IF (KEY(6)) 524, 523, 524
523 C(K,K) = C(K,K) + 1.0/VAR
    D(K) = D(K) + AVE/VAR
    GO TO 525
524 C(K,K) = C(K,K) + 1.0/(VAR*ARBI)
    D(K) = D(K) + AVE/(VAR*ARBI)
525 WRITE (6,506) I, J, AVE, VAR
C   OBTAIN POSTERIOR
526 DO 530 I = 1,NN
    B(I,1) = D(I)
    DO 530 J = 1,NN
530 A(I,J) = C(I,J)
501 FORMAT (52H1BAYESIAN ESTIMATION OF THE TRANSITION
    1PROBABILITIES//)
502 FORMAT (35H0 THE LIST OF THE PRIOR PARAMETERS/72H0 I          J
    1        R(I,J)        S(I,J)                  MEAN      VARIANCE/)
```

```
  503 FORMAT (216, 2F15.4, 2F15.8)
  504 FORMAT (72H0THE FOLLOWING XSX AND XSY DENOTE
     1XSX + S(0) AND XSY + S(0) P(0) RESPECTIVELY//)
  505 FORMAT (32H0THE LIST OF THE PRIOR KNOWLEDGE//42H I          J
     1        MEAN      VARIANCE/)
  506 FORMAT (216, 2F15.8)
  509 FORMAT (69H0THE PRIOR PARAMETERS ARE THE SAME AS
     1LISTED IN THE PREVIOUS PROBLEM./)
  540 WRITE (6,504)
      CALL GLSQP(2)
      RETURN
      END
C
C
C     THIS SUBROUTINE INITIATES THE PROBLEM CONCERNING
C     THE WEIGHT MATRIX AND THE COLUMN DROPPED TO FORM
C     THE MULTIVARIATE STRUCTURE WITH NON-SINGULAR
C     WEIGHT MATRIX.
C
      SUBROUTINE OMEGA(KO)
C
      DIMENSION A(30,30), B(30,2), C(30,30), D(30), E(30,30), X(96,30),
     1Y(96), S(96,96), XS(30,96), KEY(9), NP(96)
      COMMON A, B, C, D, E, X, Y, XS, S, SS, TOL, KDROP, KW, KV, KP,
     1KEY, LD, LR, LC, LX, MM, NN, NP, NS, NS1, NT, NT1, NOCYCL
C
      GO TO (700,743,790,800,800), KO
C     DROP A COLUMN IF REQUIRED
  700 IF (KDROP-NS) 705, 702, 701
  701 KDROP = NS
  702 DO 703 I = 1,NS
  703 NP(I) = I
      GO TO 735
  705 DO 710 I = 1,NT1
      DO 710 J = 1,NS
  710 E(I,J) = A(I,J)
      DO 720 I = 1,NT1
      K = 0
      DO 720 J = 1,NS
      IF (J-KDROP) 715, 720, 715
  715 K = K + 1
      NP(K) = J
      A(I,K) = E(I,J)
  720 CONTINUE
```

```
        DO 730 I = 1, NT1
730   A(I, NS) = E(I, KDROP)
        NP(NS) = KDROP
735   WRITE (6, 740) KDROP
        WRITE (6, 741) (NP(I), I = 1, NS)
        GO TO 744
740   FORMAT (23H0AS YOU DESIRED, COLUMN, I2, 34H IS DROPPED
      1IN FORMING THE SYSTEM.)
741   FORMAT (44H READ THE FOLLOWING MATRICES IN THE
      1SEQUENCE, 6I3/)
742   FORMAT (50H0NO COLUMN HAS BEEN DROPPED IN FORMING
      1THE SYSTEM.)
C       CHOICE OF WEIGHT
743   WRITE (6, 742)
744   NS2 = NS + 1
        DO 745 I = 1, NT1
        DO 745 J = 1, NS2
745   E(I, J) = A(I, J)
        IF (KW) 753, 753, 746
746   GO TO (755, 764, 770, 758, 760, 747, 748, 748, 750, 790), KW
747   IF (KDROP) 764, 764, 762
748   WRITE (6, 749)
749   FORMAT (65H0THE WEIGHT MATRIX IS DIAGONAL WITH
      1ELEMENTS N/(WJ(T)*(1 − WJ(T))))./)
        CALL SIGMA(5)
        GO TO 800
750   WRITE (6, 751)
751   FORMAT (37H0THIS WILL BE A TWO STAGE ESTIMATION./)
753   WRITE (6, 754)
754   FORMAT (54H0THE WEIGHT MATRIX IS AN IDENTITY MATRIX
        (UNWEIGHTED)./)
        CALL SIGMA(1)
        GO TO 800
755   IF (KDROP) 764, 764, 756
756   WRITE (6, 757)
757   FORMAT (116H0THE WEIGHT MATRIX IS DERIVED FROM THE
      1MLE WITH ELEMENTS  N/WR(T) + Z*(N/WJ(T)),  Z = 1 FOR
      2DIAGONALS AND  Z = 0 OTHERWISE./)
        CALL SIGMA(7)
        GO TO 800
758   WRITE (6, 759)
759   FORMAT (72H0THE WEIGHT MATRIX IS DIAGONAL WITH THE
      1INVERSE OF THE MEAN PROPORTIONS./)
        CALL SIGMA(2)
```

```
       GO TO 800
  760  WRITE (6,761)
  761  FORMAT (87H0THE WEIGHT MATRIX IS DIAGONAL WITH THE
      1INVERSE OF THE PRODUCT OF THE MEAN PROPORTIONS./)
       CALL SIGMA(3)
       GO TO 800
  762  WRITE (6,763)
  763  FORMAT (111H0THE WEIGHT MATRIX IS DERIVED FROM THE
      1GENERALIZED INVERSE. ONE COLUMN IS DROPPED TO PRE-
      2VENT MULTICOLINEARITY./)
       CALL SIGMA(8)
       GO TO 800
  764  WRITE (6,765)
  765  FORMAT (53H0THE WEIGHT MATRIX IS DIAGONAL WITH
      1ELEMENTS N/WJ(T)./)
       CALL SIGMA(4)
       GO TO 800
  770  READ 771, TESTA, NUMBER
  771  FORMAT (A6, 6X, I6)
       DATA WEIGHT/'WEIG'/
       IF (TESTA-WEIGHT) 772, 774, 772
  772  WRITE (6,773)
  773  FORMAT (22H0ERROR IN INPUT CARDS./)
       CALL EXIT
  774  CALL SIGMA(9)
       DO 778 K = 1, NUMBER
       READ (5,776) I, J, WEIGH
  776  FORMAT (2I6, F15.4)
  778  S(I,J) = WEIGH
       WRITE (6,779)
  779  FORMAT (43H0THE WEIGHT MATRIX IS ASSIGNED BY THE
      1USER./)
       IF (KEY(2) − 1) 800, 780, 800
  780  WRITE (6,784)
  784  FORMAT (26H0THE ASSIGNED WEIGH MATRIX/)
       CALL TABLE (1,MM,2)
       GO TO 800
  790  WRITE (6,791)
  791  FORMAT (115H0THE WEIGHT MATRIX IS DIAGONAL WITH
      1ELEMENTS THE INVERSE OF THE ESTIMATE OF THE I-TH
      2EQUATION DISTURBANCE VARIANCE./)
       CALL SIGMA(6)
C      IN ORDER TO REDUCE RANGES BETWEEN COUNTER PARTS,
C      THE WEIGH MATRIX WILL BE DIVIDED BY THE NUMBER AS-
```

```
C      SIGNED.
  800 IF (KEY(6)) 805, 820, 805
  805 ARBI = 10.0**KEY(6)
      DO 810 I = 1,MM
      DO 810 J = 1,MM
  810 S(I,J) = S(I,J)/ARBI
      WRITE (6,815) ARBI
  815 FORMAT (48HTHE WEIGHT MATRIX WILL BE DIVIDED BY
     1THE SCALAR, E10.2)
  820 IF (KEY(1) − 1) 825, 826, 826
  825 SS = 1.0
  826 IF (KEY(8) − 1) 835, 835, 836
  836 IF (KO–5) 830, 827, 830
  827 CALL GLSQP(5)
      RETURN
  830 CALL GLSQP(1)
  835 RETURN
      END
C
C
C      THIS SUBROUTINE FORMS THE MATRICES S, X, AND Y FOR THE
C      MULTIVARIATE STRUCTURE.
C
       SUBROUTINE SIGMA(MAP)
C
      DIMENSION A(30,30), B(30,2), C(30,30), D(30), E(30,30), X(96,30),
     1Y(96), S(96,96), XS(30,69), KEY(9), NP(96)
      COMMON A, B, C, D, E, X, Y, XS, S, SS, TOL, KDROP, KW, KV, KP,
     1KEY, LD, LR, LC, LX, MM, NN, NP, NS, NS1, NT, NT1, NOCYCL
C
      TN = NT
      SN = NS
      DO 100 I = 1, MM
      DO 100 J = 1, MM
  100 S(I,J) = 0.0
      GO TO (120, 130, 130, 160, 160, 170, 210, 210, 300, 350), MAP
C      UNWEIGHTED
  120 DO 121 K = 1, MM
  121 S(K,K) = 1.0
      GO TO 300
C      HOMOSKEDASTICITY WITHIN  EQUATIONS BUT HETEROSKE-
C      DASTICITY AMONG EQUATIONS
  130 DO 131 J = 1, NS1
  131 X(1,J) = 0.0
```

```
          DO 136 J = 1, NS1
          DO 135 I = 2, NT1
     135  X(1,J) = X(1,J) + A(I,J)
     136  X(1,J) = TN/X(1,J)
          IF (MAP-2) 120, 140, 150
C         HETEROSKEDASTICITY BY THE INVERSE OF THE MEAN
C         PROPORTIONS
     140  DO 145 J = 1, NS1
          LL = 1 + NT*(J - 1)
          LS = LL + NT - 1
          DO 145 K = LL, LS
     145  S(K,K) = X(1,J)
          GO TO 290
C         HETEROSKEDASTICITY BY THE INVERSE OF THE PRODUCT
C         OF THE AVERAGE PROPORTION IN STATE I AND THE AVER-
C         AGE PROPORTION NOT IN STATE I
     150  DO 155 J = 1, NS1
          SIGMA1 = X(1,J)*X(1,J)/(X(1,J) - 1.0)
          LL = 1 + NT*(J - 1)
          LS = LL + NT - 1
          DO 155 K = LL, LS
     155  S(K,K) = SIGMA1
          GO TO 290
C         DIAGONAL HETEROSKEDASTICITY BY N/WJ(T) OR BY N/(WJ
C         (T)*(1 - WJ(T)))
     160  K = 0
          DO 166 J = 1, NS1
          DO 166 I = 2, NT1
          K = K + 1
          IF (A(I,J)) 162, 162, 161
     161  IF (A(I,J) - 1.0) 163, 162, 163
     162  S(K,K) = 10.0*E(I,NS + 1)*E(I,NS + 1)
          GO TO 166
     163  IF (MAP - 4) 164, 164, 165
     164  S(K,K) = E(I,NS + 1)/A(I,J)
          GO TO 166
     165  S(K,K) = E(I,NS + 1)/(A(I,J)*(1.0 - A(I,J)))
     166  CONTINUE
          GO TO 290
C         HETEROSKEDASTICITY BY THE INVERSE OF THE I-TH EQUA-
C         TION DIST. VARIANCE.
     170  DO 171 J = 1, NS1
     171  X(1,J) = (TN - SN)/X(1,J)
          IF (KEY(2) - 2) 140, 172, 140
```

```
      172 KEY(2) = 1
          GO TO 140
C         THE WEIGHT DERIVED FROM THE MAXIMUM LIKELIHOOD
C         FUNCTION
      210 DO 280 I = 1, MM
          DO 280 L = 1, NS1
C         J IS DETERMINED BY L, AND I MINUS THE MULTIPLE OF NT
          IR = I
      215 IF (IR − NT) 225, 225, 220
      220 IR = IR − NT
          GO TO 215
      225 J = IR + NT*(L − 1)
C         K IS DETERMINED BY I MINUS THE MULTIPLE OF NT
          K = IR
C         CHECK WHETHER IT IS A DIAGONAL ELEMENT
          IF (I − J) 245, 250, 245
      245 Z = 0.0
          GO TO 255
      250 Z = 1.0
      255 CONTINUE
C         STORE THE APPROPRIATE VALUE. IF A IS ZERO, REPLACED
C         BY 0.1/SS
          SIGMA1 = A(K + 1, NS)
          SIGMA2 = A(K + 1, L)
          IF (SIGMA1) 256, 256, 257
      256 SIGMA1 = 0.1/E(K + 1, NS + 1)
      257 IF (SIGMA2) 258, 258, 259
      258 SIGMA2 = 0.1/E(K + 1, NS + 1)
      259 IF (MAP − 7) 265, 260, 265
      260 S(I,J) = (E(K + 1, NS + 1)/SIGMA1) + Z*(E(K + 1, NS + 1)/
         1SIGMA2)
          GO TO 280
      265 S(I,J) = (E(K + 1, NS + 1)/(SIGMA1*SN) + Z*(E(K + 1, NS + 1)/
         1SIGMA2) − E(K + 1, NS + 1)/(SIGMA2*SN)
      280 CONTINUE
C         FOR PROGRAMMING CHECKING–OPTION
      290 IF (KEY(2) − 1) 300, 294, 300
      294 WRITE (6,295)
      295 FORMAT (18H0THE WEIGH MATRIX/)
          CALL TABLE (1, MM, 2)
C         MVS–FORMING A KRONECKER PRODUCT
      300 NN = NS*NS1
          DO 310 I = 1, MM
          DO 310 J = 1, NN
```

```
    310  X(I,J) = 0.0
         I = 0
         DO 320 KT = 1, NS1
         DO 320 K = 1, NT
         I = I + 1
         J = (KT − 1)*NS
         DO 320 L = 1, NS
         J = J + 1
    320  X(I, J) = E(K,L)
         I = 0
         DO 330 L = 1, NS1
         DO 330 K = 2, NT1
         I = I + 1
    330  Y(I) = E(K,L)
    350  RETURN
         END
C
C
C      GENERALIZED LEAST SQUARES AND QUADRATIC PROGRAM-
C      MING.
C
       SUBROUTINE GLSQP(NCASE)
C
       DIMENSION  A(30,30),  B(30,2),  C(30,30),  D(30),  E(30,30),  X(96,30),
      1Y(96), S(96,96), XS(30,96), KEY(9), NP(96)
       COMMON A, B, C, D, E, X, Y, XS, S, SS, TOL, KDROP, KW, KV, KP,
      1KEY, LD, LR, LC, LX, MM, NN, NP, NS, NS1, NT, NT1, NOCYCL
C
       GO TO (400, 420, 616, 625, 400), NCASE
    400  DO 410 I = 1, NN
         DO 410 J = 1, MM
         XS(I,J) = 0.0
         DO 410 K = 1, MM
    410  XS (I,J) = XS(I,J) + X(K,I)*S(K,J)
         DO 416 I = 1, NN
         B(I,1) = 0.0
         DO 416 L = 1, MM
    415  B(I,1) = B(I,1) + XS(I,L)*Y(L)
         DO 416 J = 1, NN
         A(I,J) = 0.0
         DO 416 K = 1, MM
         A(I,J) = A(I,J) + XS(I,K)*X(K,J)
    416  CONTINUE
    420  DO 421 I = 1, NN
```

```
      D(I) = B(I,1)
      DO 421 J = 1, NN
  421 C(I,J) = A(I,J)
      IF (NCASE–5) 417, 700, 417
  417 IF (KEY(8)–NCASE–8) 422, 700, 422
  422 IF (KEY(3)–1) 426, 424, 423
  423 IF (KEY(3)–NCASE–1) 426, 424, 426
  424 WRITE (6,490)
      CALL PRINT (1, 1, NN, NN)
      WRITE (6,491)
      DO 425 I = 1, NN
  425 WRITE (6,495) I, B(I,1)
  426 IF (KEY(5) – 3) 427, 428, 428
  427 CALL SWPMAT (NN, 1, TOL, DETERM, 0)
      IF (DETERM) 461, 480, 461
  428 CALL SWPMAT (NN, 1, TOL, DETERM, 1)
      IF (DETERM) 460, 480, 460
  460 WRITE (6,494) DETERM
C     PRINTING P AND REGISTERING NEGATIVE P
  461 NEG = 0
      KL = NN
      WRITE (6,492)
      DO 470 K = 1, NS
      L = 0
      PR = 1.0
      N = NN – NS + K
      DO 465 I = K, N, NS
      L = L + 1
      Y(L) = B(I,1)
      IF (B(I,1)) 464, 465, 465
  464 NEG = NEG + 1
  465 PR = PR – B(I,1)
      KL = KL + 1
      B(KL,1) = PR
      IF (PR) 466, 467, 467
  466 NEG = NEG + 1
  467 CONTINUE
      IF (KDROP) 468, 468, 469
  468 WRITE (6,496) (Y(L), L = 1, NS1)
      GO TO 470
  469 WRITE (6,496) (Y(L), L = 1, NS1), PR
  470 CONTINUE
      IF (KEY(9) – 1) 472, 471, 472
  471 CALL SMRY(2,1)
```

```
472 IF (KV-1) 481, 474, 473
473 IF (KV-NCASE-1) 481, 474, 475
474 WRITE (6,488)
    CALL PRINT (1, 1, NN, NN)
    GO TO 481
475 IF (NEG) 472, 472, 481
480 WRITE (6,493)
    GO TO 530
481 IF (KP-1) 487, 486, 482
482 IF (KP-NCASE-1) 487, 486, 483
483 IF (NEG) 487, 486, 487
486 CALL PREDIC (1)
487 IF (KEY(7) - 1) 700, 501, 501
488 FORMAT (102H0THE DISPERSION MATRIX OF THE ML (GLS,
    1MCS) ESTIMATOR P(I,J), OR THE INVERSE OF XSX FOR ANY-
    2THING ELSE./)
489 FORMAT (39H0THE UNRESTRICTED ESTIMATOR IS PERFECT./)
490 FORMAT (29H0GENERALIZED XSX FOR QP INPUT/)
491 FORMAT (29H0GENERALIZED XSY FOR QP INPUT/)
492 FORMAT (48H0UNRESTRICTED ESTIMATOR OF THE TRANSI-
    1TION MATRIX/)
493 FORMAT (33H0XSX IS SINGULAR, TRY QP SOLUTION/)
494 FORMAT (21H0DETERMINANT OF XSX =, E14.7)
495 FORMAT (1H, I3, 3X, E14.7)
496 FORMAT (8F15.8)
497 FORMAT (67H0THE UNRESTRICTED ESTIMATOR VIOLATES
    1THE PROPERTIES OF PROBABILITY. /44H HENCE, THE ITERA-
    2TION PROCEDURE IS REQUIRED./)
C       IF P IS NOT IN THE RANGE BETWEEN ONE AND ZERO CALL
C       FOR QP SOLUTION
501 IF (KEY(7)-NCASE) 700, 510, 502
502 IF (KEY(7)-3) 700, 510, 510
510 IF (NEG) 511, 511, 512
511 WRITE (6,489)
    GO TO 700
512 WRITE (6, 497)
    IF (KEY(7)-4) 530, 520, 520
520 K4 = KEY(4)
    IF (KEY(7)-NCASE-3) 525, 530, 525
525 KEY(4) = 0
530 CALL QP(1)
    LX = LR
    IF (KEY(7)-4) 540, 531, 531
531 KEY(4) = K4
```

C   OUTPUTING THE QP SOLUTIONS IN THE MATRIX FORM
540 IF (KW–1) 548, 542, 541
541 GO TO (542, 546, 546, 546, 546, 549, 546, 546, 548, 547), KW
542 IF (NCASE–1) 544, 544, 545
544 WRITE (6,560)
     GO TO 616
545 WRITE (6,561)
     GO TO 616
546 WRITE (6,562)
     GO TO 616
547 WRITE (6,563)
     GO TO 616
548 WRITE (6,564)
     GO TO 616
549 WRITE (6,565)
     GO TO 616
560 FORMAT (63H0HENCE, THE ML (GLS, MCS) ESTIMATOR OF THE
     1TRANSITION MATRIX IS/)
561 FORMAT (58H0HENCE, THE BAYESIAN ESTIMATOR OF THE
     1TRANSITION MATRIX IS/)
562 FORMAT (58H0HENCE, THE WEIGHTED ESTIMATOR OF THE
     1TRANSITION MATRIX IS/)
563 FORMAT (59H0HENCE, THE TWO-STAGE ESTIMATOR OF THE
     1TRANSITION MATRIX IS/)
564 FORMAT (70H0HENCE, THE UNWEIGHTED CLASSICAL ESTI-
     1MATOR OF THE TRANSITION MATRIX IS/)
565 FORMAT (92H0HENCE, THE FIRST STAGE RESTRICTED GENE-
     1RALIZED INVERSE ESTIMATOR OF THE TRANSITION MATRIX
     2IS/)
616 DO 617 I = 1, LR
617 B(I,1) = 0.0
     DO 620 I = 1, LR
     K = NP(I) − LD
     IF (K) 620, 620, 618
618 IF (K–LX) 619, 619, 620
619 B(K,1) = Y(I)
620 CONTINUE
     IF (KEY(9)–1) 625, 621, 625
621 CALL SMRY(2,2)
625 DO 650 L = 1, NS
     LL = 0
     N = NN + L
     DO 630 I = L, N, NS
     LL = LL + 1

```
      630 Y(LL) = B(I,1)
      650 WRITE (6,496) (Y(LL), LL = 1, NS)
          IF (NCASE-4) 651, 700, 651
      651 IF (KP-1) 700, 660, 652
      652 IF (KP-NCASE-1) 700, 660, 653
      653 IF (KP-4) 700, 660, 660
      660 CALL PREDIC (1)
      700 RETURN
          END
C
C
C     THIS SUBROUTINE FORMS A SIMPLEX TABLEAU FOR A LI-
C     NEAR PROGRAMMING FITTING A SET OF REGRESSION HY-
C     PERPLANES BY MINIMIZING THE SUM OF ABSOLUTE DIS-
C     TANCES, WEIGHTED OR UNWEIGHTED.
C
      SUBROUTINE LP
C
      DIMENSION A(30,30), B(30,2), C(30,30), D(30), E(30,30), X(96,30),
     1Y(96), S(96,96), XS(30,96), KEY(9), NP(96)
      COMMON A, B, C, D, E, X, Y, XS, S, SS, TOL, KDROP, KW, KV, KP,
     1KEY, LD, LR, LC, LX, MM, NN, NP, NS, NS1, NT, NT1, NOCYCL
C
      WRITE (6,100)
  100 FORMAT (75H0YOU HAVE INDICATED THAT THIS PROBLEM
     1WILL BE SOLVED BY LINEAR PROGRAMMING./)
      IF (KDROP) 101, 101, 102
  101 CALL OMEGA(2)
      GO TO 104
  102 CALL OMEGA(1)
  104 NNS = NS*NS
      NN1 = NN + 1
      MM1 = MM + 1
      LR = MM + NS
      LX = LC
      LD = 0
      NOCYCL = 0
C     FORM A SIMPLEX TABLEAU FOR LP
      DO 105 J = 1, LC
  105 XS(1,J) = 0.0
      DO 110 K = 1, MM
      J = NNS + K
      L = MM + J
      DO 106 I = 1, MM
```

```
106 XS(1,J) = XS(1,J) + S(I,K)
110 XS(1,L) = XS(1,J)
    DO 111 I = 1, MM
    DO 111 J = 1, NN
111 S(I,J) = X(I,J)
    DO 112 J = 1, LC
112 X(J,1) = XS(1,J)
    IF (KEY(8)) 113, 113, 115
113 K = NNS + 1
    DO 114 J = K, LC
114 X(J,1) = SQRT(X(J,1))
115 IF (KDROP-1) 120, 116, 116
116 DO 117 I = 1, MM
    DO 117 J = NN1, NNS
117 S(I,J) = 0.0
120 IF (KEY(4)-1) 122, 123, 122
122 IF (KEY(2)-1) 129, 123, 123
123 K = NNS + 1
    WRITE (6,124) K, LC
124 FORMAT (40H0THE COST VECTOR FOR ERROR TERMS (CO-
    1LUMN, I3, 10H TO COLUMN, I3, 1H)/)
    WRITE (6,125) (X(J,1), J = K, LC)
125 FORMAT (1H, 9E14.7)
C       FORM ERROR TERMS IN THE FORM OF COUNTER PARTS AND
C       INTRODUCE INTO BASIS
129 DO 130 I = 1, MM
    DO 130 J = NN1, LC
130 S(I,J) = 0.0
    DO 135 I = 1, MM
    J = NNS + I
    NP(I) = J
    S(I,J) = 1.0
    L = MM + J
135 S(I,L) = -1.0
C       FORM ROW SUM CONSTRAINTS
    DO 140 I = MM1, LR
    NP(I) = 0
140 Y(I) = 1.0
    DO 145 I = MM1, LR
    DO 145 J = 1, LC
145 S(I,J) = 0.0
    DO 146 K = 1, NS
    I = MM + K
    L = NNS - NS + K
```

```
        DO 146 J = K, L, NS
146   S(I,J) = 1.0
        IF (KDROP) 150, 150, 147
147   J = NN
        DO 148 I = MM1, LR
        J = J + 1
148   NP(I) = J
        IF (KEY(4)–1) 156, 149, 156
149   CALL TABLE(1,LC,1)
        GO TO 156
C       PHASE I ITERATION
150   IF (KEY(4)–1) 152, 151, 152
151   CALL TABLE (1,LC,1)
152   CALL SOLVE (1)
        IF (KEY(5)–2) 156, 154, 153
153   IF (KEY(5)–5) 156, 154, 156
154   WRITE (6,155)
155   FORMAT (42H END OF PHASE I AND BEGINNING OF PHASE II.)
C       PHASE II ITERATION
156   DO 160 I = 1, LR
        M = NP(I)
160   X(I,2) = X(M,1)
        KS = LR + 1
        DO 165 J = 1, LC
        S(KS,J) = 0.0
        DO 164 I = 1, LR
164   S(KS,J) = S(KS,J) + X(I,2)*S(I,J)
165   S(KS,J) = S(KS,J) – X(J,1)
        Y(KS) = 0.0
        DO 170 I = 1, LR
170   Y(KS) = Y(KS) + X(I,2)*Y(I)
        IF (KDROP) 172, 172, 175
172   CALL SOLVE (3)
        GO TO 176
175   CALL SOLVE (2)
176   Y(KS) = 0.0
        DO 180 I = 1, LR
        M = NP(I)
        X(I,2) = X(M,1)
180   Y(KS) = Y(KS) + X(I,2)*Y(I)
        CALL QP(2)
C       OUTPUT THE TRANSITION MATRIX
        WRITE (6,201)
201   FORMAT (66H0THE MINIMUM ABSOLUTE DEVIATION ESTI-
```

```
1MATOR OF THE TRANSITION MATRIX/)
 LX = NNS
 CALL GLSQP(3)
 RETURN
 END
C
C
 SUBROUTINE QP(KOUT)
C
C     THIS SUBROUTINE FORMS SIMPLEX TABLEAUX FOR QUADRA-
C     TIC PROGRAMMING PROBLEMS, WITH EQUALITY OR IN-
C     EQUALITY CONSTRAINTS.
 DIMENSION A(30,30), B(30,2), C(30,30), D(30), E(30,30), X(96,30),
 1Y(96), S(96,96), XS(30,96), KEY(9), NP(96)
 COMMON A, B, C, D, E, X, Y, XS, S, SS, TOL, KDROP, KW, KV, KP,
 1KEY, LD, LR, LC, LX, MM, NN, NP, NS, NS1, NT, NT1, NOCYCL
C
 GO TO (3, 100), KOUT
C     DEFINE CONSTANTS
 3 LD = NS
 LR = NS*NS
 IF (KDROP) 5, 5, 8
 5 LR = LR + NS
 8 LC = 2*LR
 LX = 2*LC
C     STORE THE VALUE FOR P-ZERO COLUMN, OR RIGHT HAND
C     SIDE.
 DO 10 I = 1, NS
 10 Y(I) = 1.0
 K = 0
 NS2 = NS + 1
 DO 15 I = NS2, LR
 K = K + 1
 15 Y(I) = D(K)
C     CLEAR THE CORE LOCATIONS FOR LEFT HAND SIDE.
 DO 20 I = 1, LR
 DO 20 J = 1, LC
 20 S(I, J) = 0.0
C     FORM G-PRIME AND G MATRICES (ALSO − G-PRIME FOR
C     EQUALITY TYPE).
 DO 30 I = 1, LD
 KR = LD + I
 KC = NN + I
 DO 30 J = KR, KC, NS
```

```
          S(I,J) = 1.0
          IF (KDROP) 25, 25, 30
       25 L = LR + I
          S(J,L) = -1.0
       30 S(J,I) = 1.0
C         FORM BETA-MATRIX
          DO 40 K = 1, NN
          I = LD + K
          DO 40 L = 1, NN
          J = LD + L
       40 S(I,J) = C(K,L)
C         FORM SLACK VARIABLES, W(I).
          IF (KDROP) 55, 55, 48
       48 DO 50 I = 1, NS
          J = LR + I
       50 S(I,J) = 1.0
       55 DO 60 I = NS2, LR
          J = LR + I
       60 S(I,J) = -1.0
C         INITIALIZATION
          DO 70 K = 1, LR
       70 NP(K) = 0
          IF (KDROP) 85, 85, 75
C         INTRODUCE ALL THE ELEMENTS OF THE DROPPED COLUMN
C         OF P INTO THE BASIS INITIALLY AS A SHORT-CUT.
       75 DO 80 K = 1, LD
       80 NP(K) = LR + K
C         REDEFINE CONSTANTS SO THAT P(I)S ARE COUNTERPARTS
C         TO W(I)S.
       85 LD = LD + NN
          NOCYCL = 0
C         OUTPUT THE INITIAL TABLEAU - OPTION
          IF (KEY(4)-1) 95, 90, 95
       90 CALL TABLE (1, LC, 1)
       95 WRITE (6,190)
          CALL SOLVE(1)
          LD = NS
C         OUTPUT ROUTINE
      100 KS = LR + 1
          WRITE (6,191) NOCYCL, Y(KS)
          IF (KEY(5)) 200, 200, 105
      105 IF (KEY(5)-3) 106, 200, 106
      106 WRITE (6,192)
          DO 110 I = 1, LR
```

```
110  WRITE (6,193) I, NP(I), Y(I)
190  FORMAT (32H1THE FOLLOWINGS ARE QP SOLUTIONS//)
191  FORMAT (28H0NUMBER  OF  CYCLES  REQUIRED = I4,  49H,
     1OBJ. VALUE FOR LP OR ROUNDING ERROR FOR QP = E15.7/)
192  FORMAT (44H0            I            J            B(I)/)
193  FORMAT (2I10, F30.10)
200  RETURN
     END
C
C
C        THIS SUBROUTINE SOLVES SIMPLEX TABLEAUX FOR LP AND
C        ALSO FOR QP. BY T.C.LEE.
C
         SUBROUTINE SOLVE (LQP)
C
         DIMENSION A(30,30), B(30,2), C(30,30), D(30), E(30,30), X(96,30),
     1Y(96), S(96,96), XS(30,96), KEY(9), NP(96)
         COMMON A, B, C, D, E, X, Y, XS, S, SS, TOL, KDROP, KW, KV, KP,
     1KEY, LD, LR, LC, LX, MM, NN, NP, NS, NS1, NT, NT1, NOCYCL
C
         ASSIGN 135 TO KK
         KS = LR + 1
         IF (LQP–2) 16, 16, 380
  16  IF (KEY(5)–2) 80, 19, 17
  17  IF (KEY(5)–5) 80, 19, 80
  19  WRITE (6,20)
  20  FORMAT (84H0NO. OF ITERATIONS PIVOT ROW NEW BASIS
     1OLD BASIS OBJ. VALUE BEFORE ITERATION)
  80  IF (LQP–1) 90, 90, 380
  90  NOCYCL = 0
C        OBTAIN ALTERNATIVE COSTS
 100  LOOP = 0
         DO 110 J = 1, LC
 110  S(KS,J) = 0.0
         DO 120 J = 1, LC
         DO 120 I = 1, LR
         IF (NP(I)) 120, 119, 120
 119  S(KS,J) = S(KS,J) + S(I,J)
 120  CONTINUE
         Y(KS) = 0.0
         DO 125 I = 1, LR
         IF (NP(I)) 125, 124, 125
 124  Y(KS) = Y(KS) + Y(I)
 125  CONTINUE
```

```
C       SMALL PROBLEM MAY WISH TO KNOW ITERATION STEPS
    131 IF (KEY(4)–9) 135, 132, 135
    132 CALL TABLE (1, LC, 1)
        GO TO KK, (135,400)
C       CHECK WHETHER TO GO OR TO STOP
    135 IF (NOCYCL–LX) 140, 136, 400
    136 WRITE (6,137) Y(KS)
    137 FORMAT (82H0I THINK I AM LOOPING, SO I WILL TERMINATE.
        1THE VALUE OF THE OBJECTIVE FUNCTION =, E20.10/)
    138 IF (KEY(4)–8) 400, 139, 400
    139 NOCYCL = LX + 1
        GO TO 132
C       FIND BASIS AMONG ACTIVE COLUMNS
    140 KMAX = 0
        DO 160 J = 1, LC
        IF (KMAX) 501, 501, 500
    500 IF (S(KS,J)) 160, 160, 143
    501 IF (S(KS,J)) 160, 141, 143
    141 DO 142 I = 1, LR
        IF (NP(I)–J) 142, 160, 142
    142 CONTINUE
    143 IF (LD) 150, 150, 144
    144 IF (J–LD) 147, 147, 145
    145 IF (J–LR) 150, 150, 146
    146 L = J – LR
        GO TO 148
    147 L = J + LR
    148 DO 149 K = 1, LR
        IF (NP(K)–L) 149, 160, 149
    149 CONTINUE
    150 IF (KMAX) 151, 151, 152
    151 KMAX = J
        GO TO 160
    152 IF(S(KS,KMAX)–S(KS,J)) 153, 160, 160
    153 KMAX = J
    160 CONTINUE
        IF (KMAX) 161, 161, 210
    161 IF (LD) 235, 235, 172
C       TAKE CARE OF LOOPING OR LINEAR DEPENDENCY
    165 DO 169 I = 1, LR
        IF (NP(I)) 169, 167, 169
    167 SUM = 0.0
        DO 168 J = 1, LC
    168 SUM = SUM + ABS(S(I,J))
```

```
      IF (SUM) 169, 170, 169
  169 CONTINUE
      GO TO 100
  170 WRITE (6,171) I
  171 FORMAT (88H FORCED TERMINATION DUE TO LINEAR DE-
     1PENDENCE OR INCONSISTENCY OF CONSTRAINTS. CHECK
     2ROW, I5)
      GO TO 138
C     ERASE THE COUNTER PARTS OF THE POSSIBLE NEW BASES
C     FROM THE CURRENT BASES.
  172 WRITE (6,173)
  173 FORMAT (54H I CAN NOT FIND A BASIS AND WILL NOW ALTER
     1ACTIVITIES.)
      KAN = 0
      DO 181 I = 1, LR
      IF (NP(I)) 181, 174, 181
  174 DO 181 J = 1, LC
      IF (S(I,J)) 181, 181, 175
  175 IF (J–LD) 177, 177, 176
  176 IF (J–LR) 181, 181, 178
  177 L = J + LR
      GO TO 179
  178 L = J – LR
  179 DO 181 K = 1, LR
      IF (NP(K)–L) 181, 180, 181
  180 KAN = KAN + 1
      NP(K) = 0
  181 CONTINUE
      IF (KAN) 235, 235, 182
  182 WRITE (6,183) KAN
  183 FORMAT (10H THERE ARE, I6, 14H BASES ERASED.)
      GO TO 100
C     FIND PIVOT RATIOS
  210 DO 220 I = 1, LR
      IF (S(I,KMAX)) 215, 215, 212
  212 XS(1,I) = Y(I)/S(I,KMAX)
      GO TO 220
  215 XS(1,I) = −1.0
  220 CONTINUE
C     FIND THE SMALLEST NONNEGATIVE PIVOT RATIO
      KPIVOT = 0
      DO 225 I = 1, LR
      IF (XS(1,I)) 225, 221, 221
  221 IF (KPIVOT) 222, 222, 223
```

```
      222 KPIVOT = I
          GO TO 225
      223 IF (XS(1,KPIVOT)-XS(1,I)) 225, 225, 224
      224 KPIVOT = I
      225 CONTINUE
          IF (KPIVOT) 235, 235, 248
C         IF NO NONNEGATIVE PIVOT RATIO, NO FEASIBLE SOLUTION
C         EXISTS
      235 WRITE (6,236)
      236 FORMAT (39H0THIS PROBLEM HAS NO FEASIBLE SOLUTION./)
          GO TO 138
C         RECORD POSSIBLE LOOPING AND PRINT LOCATION OF BASIS
      248 IF (LQP-1) 249, 249, 260
      249 IF (NP(KPIVOT)) 255, 255, 250
      250 LOOP = LOOP + 1
          IF (LOOP-9) 260, 260, 251
      251 WRITE (6,252)
      252 FORMAT (62H I AM CHECKING FOR LOOPING, AND RECOM-
         1PUTING ALTERNATIVE COSTS.)
          GO TO 165
      255 LOOP = 0
      260 NOCYCL = NOCYCL + 1
          IF (KEY(5)-2) 275, 265, 262
      262 IF (KEY(5)-5) 275, 265, 275
      265 WRITE (6,270) NOCYCL, KPIVOT, KMAX, NP(KPIVOT), Y(KS)
      270 FORMAT (2H, 4I12, 10X, E14.7)
      275 NP(KPIVOT) = KMAX
C         ITERATION THROUGH MATRIX AND ALTERNATIVE COSTS
          PIVOT = 1.0/S(KPIVOT, KMAX)
          DO 340 J = 1, LC
      340 S(KPIVOT,J) = S(KPIVOT,J)*PIVOT
          Y(KPIVOT) = Y(KPIVOT)*PIVOT
          S(KPIVOT,KMAX) = 1.0
          DO 360 I = 1, KS
          IF (I-KPIVOT) 345, 360, 345
      345 IF (S(I,KMAX)) 346, 360, 346
      346 ELEMT = S(I,KMAX)
          DO 350 J = 1, LC
          IF (S(KPIVOT,J)) 347, 350, 347
      347 S(I,J) = S(I,J) - ELEMT*S(KPIVOT,J)
          IF (ABS(S(I,J))-TOL) 349, 349, 350
      349 S(I,J) = 0.0
      350 CONTINUE
          Y(I) = Y(I) - ELEMT*Y(KPIVOT)
```

```
      IF (ABS(Y(I))–TOL) 353, 353, 354
353   Y(I) = 0.0
354   S(I,KMAX) = 0.0
360   CONTINUE
C     FIND OUT WHETHER BASES ARE FILLED
      IF (LQP–1) 365, 365, 380
365   DO 370 I = 1, LR
      IF (NP(I)) 131, 131, 370
370   CONTINUE
      IF (KEY(4)–7) 371, 375, 371
371   IF (KEY(4)–9) 400, 375, 400
375   ASSIGN 400 TO KK
      GO TO 132
380   DO 385 J = 1, LC
      IF (S(KS,J)) 385, 385, 131
385   CONTINUE
C     CHECKING FOR MULTIPLE SOLUTIONS
      DO 390 J = 1, LC
      IF (S(KS,J)) 390, 386, 390
386   DO 387 I = 1, LR
      IF (NP(I)–J) 387, 390, 387
387   CONTINUE
      DO 388 K = 1, LR
      IF (S(K,J)) 388, 388, 394
388   CONTINUE
390   CONTINUE
      GO TO 370
394   WRITE (6,395) J
395   FORMAT (69H0THIS PROBLEM HAS MULTIPLE SOLUTIONS. AT
     1LEAST THE VARIABLE OF COLUMN, I4, 24H COULD BE IN THE
     2BASIS./)
      GO TO 138
400   RETURN
      END
C
C

      SUBROUTINE TABLE (K, L, LINE)
C     THIS SUBROUTINE PRINTS SIMPLEX TABLEAU AT ANY STAGE.
      DIMENSION A(30,30), B(30,2), C(30,30), D(30), E(30,30), X(96,30),
     1Y(96), S(96,96), XS(30,96), KEY(9), NP(96)
      COMMON A, B, C, D, E, X, Y, XS, S, SS, TOL, KDROP, KW, KV, KP,
     1KEY, LD, LR, LC, LX, MM, NN, NP, NS, NS1, NT, NT1, NOCYCL
C
      KS = L
```

```
      NE = K
      GO TO (409, 510), LINE
  409 WRITE (6,500) NOCYCL
  499 FORMAT (10H   Z-C,        9E13.7)
  500 FORMAT (16H0SIMPLEX TABLEAU, I6)
      KS = LR
      IF (LC-8) 501, 501, 502
  501 LS = LC
      GO TO 503
  502 LS = 8
  503 WRITE (6, 504) (J, J = 1, LS)
  504 FORMAT (18H0ROW BASIS B(I), 8I13)
  505 FORMAT (1H, I3, I5, 1X, 9E13.7)
  506 FORMAT (1H, I3, 6X, 9E13.7)
  507 FORMAT (1H0)
      DO 508 I = 1, KS
  508 WRITE (6, 505) I, NP(I), Y(I), (S(I,J), J = 1, LS)
      IF (KEY(4)-1) 600, 610, 600
  600 WRITE (6,499) Y(LR + 1), (S(LR + 1,J), J = 1, LS)
  610 WRITE (6, 507)
      IF (LC-8) 540, 540, 509
  509 NE = 9
C
  510 ASSIGN 512 TO KK
      LL = NE
  512 IF ((L-LL)-8) 515, 515, 520
  515 ASSIGN 540 TO KK
      LS = L
      GO TO 525
  520 LS = LL + 8
  525 WRITE (6,526) (J, J = LL,LS)
  526 FORMAT (1H, 5X, 9I13)
      DO 530 I = 1, KS
  530 WRITE (6,506) I, (S(I,J), J = LL, LS)
      IF (LINE-1) 531, 531, 533
  531 IF (KEY(4)-1) 533, 533, 532
  532 WRITE (6,499) (S(LR + 1, J), J = LL, LS)
  533 WRITE (6,507)
      LL = LL + 9
      GO TO KK, (512,540)
  540 RETURN
      END
C
C
```

```
      SUBROUTINE SWPMAT (N, M, TOL, DETERM, KASE)
C
C
C     THIS SUBROUTINE INVERTS THE MATRIX A AND SOLVES THE
C     EQUATION AX = B. THE ITERATIVE METHOD SEE B.W.ARDEN,
C     AN INTRODUCTION TO DIGITAL COMPUTING, PP.215-223.
C
      DIMENSION A(30,30), B(30,2)
      COMMON A, B
C
      KOVER = KASE
      DETERM = 1.0
      DO 60 K = 1, N
C     TEST FOR SIGNIFICANCE OF PIVOT ELEMENT.
      IF (ABS(A(K,K))-TOL) 80, 80, 10
   10 IF (KOVER-1) 19, 15, 19
   15 DETERM = DETERM*A(K,K)
   19 PIVOT = 1.0/A(K,K)
C     SCALE KTH ROW.
      DO 20 J = 1, N
   20 A(K,J) = A(K,J)*PIVOT
      DO 25 J = 1, M
   25 B(K,J) = B(K,J)*PIVOT
      A(K,K) = PIVOT
C     REDUCE ALL ROWS BUT KTH ROW, ALL ELEMENTS.
      DO 50 I = 1, N
      IF (I-K) 30, 50, 30
C     AUTOMATICALLY REPLACES A(I,K) WITH - A(I,K)*A(K,K)
   30 ELEMT = A(I,K)
      A(I,K) = 0.0
      DO 40 J = 1, N
   40 A(I,J) = A(I,J) - ELEMT*A(K,J)
      DO 45 J = 1, M
   45 B(I,J) = B(I,J) - ELEMT*B(K,J)
   50 CONTINUE
   60 CONTINUE
   99 RETURN
   80 WRITE (6,90) K
   90 FORMAT (31H0THE MATRIX IS SINGULAR, ON K = , I8/)
      DETERM = 0.0
      GO TO 99
      END
C
C
```

```
      SUBROUTINE PRINT (K, L, M, N)
C     THIS SUBROUTINE PRINTS MATRIX A WITH INDICES.
      DIMENSION A(30,30)
      COMMON A
C
   05 FORMAT (1H0)
   20 FORMAT (1H , 9I14)
   30 FORMAT (1H , I3, 2X, 9E14.7)
      INT = 9
      INT1 = INT − 1
   50 ASSIGN 60 TO KK
      LL = L
   60 IF ((N–LL)–INT1) 70, 70, 75
   70 ASSIGN 180 TO KK
      LS = N
      GO TO 80
   75 LS = LL + INT1
   80 WRITE (6,20) (J, J = LL, LS)
      DO 90 I = K, M
   90 WRITE (6,30) I, (A(I,J), J = LL, LS)
      WRITE (6,5)
      LL = LL + INT
      GO TO KK, (60,180)
  180 RETURN
      END
C
C
C     THIS SUBROUTINE PREDICTS THE FUTURE OUTCOMES AND
C     EVALUATES THE SQUARED ERROR AS THE VALUE OF THE
C     LOSS FUNCTION AND ALSO COMPUTES THE CHI-SQUARE
C     VALUE FOR THE TEST OF THE GOODNESS-OF-FIT.
C
      SUBROUTINE PREDIC (KN)
C
      DIMENSION A(30,30), B(30,2), C(30,30), D(30), E(30,30), X(96,30),
     1Y(96), S(96,96), XS(30,96), KEY(9), NP(96)
      COMMON A, B, C, D, E, X, Y, XS, S, SS, TOL, KDROP, KW, KV, KP,
     1KEY, LD, LR, LC, LX, MM, NN, NP, NS, NS1, NT, NT1, NOCYCL
C
      DO 20 I = 1, NT1
      L = 0
      DO 20 J = 1, NS
      A(I + 1, J) = 0.0
      DO 20 K = 1, NS
```

```
       L = L + 1
   20  A(I + 1, J) = A(I + 1, J) + E(I, K)*B(L, 1)
       IF (KN–10) 24, 60, 24
   24  WRITE (6, 25)
   25  FORMAT (26H1THE PREDICTED PROPORTIONS/)
       NT2 = NT1 + 1
       CALL PRINT (2, 1, NT2, NS)
       SSQ = 0.0
       CHI = 0.0
       DO 40 J = 1, NS
       DO 40 I = 2, NT1
       IF (A(I, J)) 38, 35, 38
   35  IF (E(I, J)) 39, 40, 39
   38  CHI = CHI + (E(I, NS + 1)*(A(I, J) − E(I, J))**2)/A(I, J)
       GO TO 40
   39  CHI = CHI + (E(I, NS + 1)*(A(I, J) − E(I, J))**2)/E(I, J)
   40  SSQ = SSQ + (A(I, J) − E(I, J))**2
       TN = NT*NS
       ASSQ = SSQ/TN
       CHIR = 0.0
       DO 45 J = 1, NS
       DO 45 I = 2, NT1
       IF (E(I, J)) 43, 42, 43
   42  IF (A(I, J)) 44, 45, 44
   43  CHIR = CHIR + (E(I, NS + 1)*(A(I, J) − E(I, J))**2)/E(I, J)
       GO TO 45
   44  CHIR = CHIR + (E(I, NS + 1)*(A(I, J) − E(I, J))**2)/A(I, J)
   45  CONTINUE
       WRITE (6, 50) SSQ, ASSQ, CHI, CHIR
   50  FORMAT (91H0 SUM OF  SQUARED  ERROR  MEAN  SQUARED
      1ERROR CHI SQUARE VALUE MODIFIED CHI SQUARE/7X, 4F20.9)
   60  RETURN
       END
C
C

       SUBROUTINE SMRY (MARY, KT)
C      THIS SUBROUTINE SUMMARIZES THE ESTIMATES FOR THE
C      SAMPLING EXPERIMENT ONLY.
       DIMENSION A(30,30), B(30,2), C(30,30), D(30), E(30,30), X(96,30),
      1Y(96), S(96,96), XS(30,96), KEY(9), NP(96), AA(2), SUM(36,2), SUMSQ
      2(36,2)
       COMMON A, B, C, D, E, X, Y, XS, S, SS, TOL, KDROP, KW, KV, KP,
      1KEY, LD, LR, LC, LX, MM, NN, NP, NS, NS1, NT, NT1, NOCYCL
C
```

```
          NNS = NS*NS
          GO TO (10, 25, 35), MARY
     10   DO 20 J = 1, 2
          AA(J) = 0.0
          DO 20 I = 1, NNS
          SUM(I, J) = 0.0
     20   SUMSQ(I, J) = 0.0
          GO TO 100
     25   DO 30 I = 1, NNS
          SUM(I, KT) = SUM(I, KT) + B(I, 1)
     30   SUMSQ(I, KT) = SUMSQ(I, KT) + B(I, 1)**2
          AA(KT) = AA(KT) + 1.0
          WRITE (7,31) (S(I,1), I = 1, NNS)
     31   FORMAT (8X, 8F8.4)
          GO TO 100
     35   IF (KT − 1) 37, 37, 39
     37   WRITE (6,38)
     38   FORMAT (56H1THIS IS A SUMMARY SHEET FOR THE UNRE-
          1STRICTED ESTIMATOR./)
          GO TO 41
     39   WRITE (6,40)
     40   FORMAT (54H1THIS IS A SUMMARY SHEET FOR THE RE-
          1STRICTED ESTIMATOR./)
     41   DO 42 I = 1, NNS
     42   B(I,1) = SUM(I,KT)/AA(KT)
          WRITE (6,44) AA(KT)
     44   FORMAT (14H0SAMPLE SIZE = F4.0)
          WRITE (6,45)
     45   FORMAT (6H0MEANS/)
          CALL GLSQP (4, TOL, KW, KV, KDROP)
          DO 47 I = 1, NNS
     47   B(I,1) = SUMSQ(I,KT)/AA(KT) − B(I,1)**2
          WRITE (6,48)
     48   FORMAT (10H0VARIANCES/)
          CALL GLSQP(4)
          DO 49 I = 1, NNS
     49   B(I,1) = SQRT(B(I,1))
          WRITE (6,50)
     50   FORMAT (20H0STANDARD DEVIATIONS/)
          CALL GLSQP(4)
     55   READ (5,56) (B(I,1), I = 1, 36)
     56   FORMAT (9F8.4/9F8.4/9F8.4/9F8.4)
          WRITE (6,57)
     57   FORMAT (12H0TRUE VALUES/)
```

```
     CALL GLSQP(4)
     DO 70 I = 1, NNS
60   Y(I) = SUMSQ(I,KT) − 2.0*B(I,1)*SUM(I,KT) + AA(KT)*B(I,1)**2
70   B(I,1) = SQRT(Y(I)/AA(KT))
     WRITE (6,75)
75   FORMAT (41H0ROOT MEAN SQUARE ERRORS FROM TRUE
     1VALUES/)
80   CALL GLSQP(4)
     S1 = 0.0
     DO 85 I = 1, NNS
85   S1 = S1 + B(I,1)
     WRITE (6,89) S1
89   FORMAT (59H0SUM OF THE ELEMENTS OF THE ROOT MEAN
     1SQUARE ERROR MATRIX =, F15.8)
100  RETURN
     END
```

REFERENCES

The published and unpublished work contained in this reference list, while not exhaustive, does contain much of the literature important to the methods developed in this book. A few of the references are not referred to in the book but are given to provide a breadth of coverage to the topics discussed.

ADELMAN, I.G., 1958, "A Stochastic Analysis of the Size Distribution of Farms" *Journal of the American Statistical Association, 53*, 893–904.

AITKEN, A.C., 1934–35, "On Least Squares and Linear Combination of Observations", *Proceedings of the Royal Society of Edinburgh, 55*, 42–48.

ANDERSON, T.W., 1955, "Probability Models for Analyzing Time Changes and Attitudes", in: P.F.Lazarsfeld, ed., *Mathematical Thinking in the Social Sciences* (The Free Press, Glencoe) pp.17–66.

ANDERSON, T.W. and L.A.GOODMAN, 1957, "Statistical Inference About Markov Chains", *The Annals of Mathematical Statistics, 28*, 89–110.

ARROW, K.J., 1967, "Applications of Control Theory to Economic Growth", Technical Report No.2, Institute of Mathematical Studies, Stanford.

ASHAR, V.G. and T.D.WALLACE, 1963, "A Sampling Study of Minimum Absolute Deviations Estimators", *Operations Research, 11*, 747–758.

BAILEY, N.T.J., 1964, The Elements of Stochastic Processes with Applications to the Natural Sciences (John Wiley and Sons, New York).

BARTEN, A.P. and T.KLOEK, 1965, "A Generalization of Generalized Least Squares", Econometric Institute Report 6508, Netherlands School of Economics.

BARTLETT, M.S., 1951, "The Frequency Goodness-of-Fit Test for Probability Chains", *Proceedings of the Cambridge Philosophical Society, 47*, 86–95.

BARTLETT, M.S., 1955, *An Introduction to Stochastic Processes* (Cambridge University Press, Cambridge) pp.24–26; 240–242.

BEALE, E.M.L., 1955, "On Minimizing a Convex Function Subject to Linear Inequalities", *Journal of the Royal Statistical Society* (B), *17*, 173–184.

BLACKWELL, D. and M.A.GIRSCHICK, 1954, *Theory of Games and Statistical Decision* (John Wiley and Sons, New York) pp.143–178.

BLUMEN, I., M.KOGAN and P.J.MCCARTHY, 1955, *The Industrial Mobility of Labor as a Probability Process* (Cornell University Press, Ithaca).

BOOT, J.C.G., 1963, "The Computation of the Generalized Inverse of Singular or Rectangular Matrices", *American Mathematical Monthly, 70*, 302–303.

BUSH, R.R. and F.MOSTELLER, 1955, *Stochastic Models for Learning* (John Wiley and Sons, New York) Chapters 3 and 5.

CHAMPERNOWNE, D.B., 1953, "A Model of Income Distribution", *Economic Journal*, *63*, 318–351.

CHARNES, A. and W.W.COOPER, 1961, *Management Models and Industrial Applications of Linear Programming*, Vol. 1 (John Wiley and Sons, New York) pp. 389 to 393.

CHIPMAN, J.C., 1964, "On Least Squares with Insufficient Observations", *Journal of the American Statistical Association*, *59*, 1078–1111.

CLINE, R.E., 1964, "Note on the Generalized Inverse of the Product of Matrices", *SIAM Review*, *6*, 57–58.

COLEMAN, J.S., 1964, *Introduction to Mathematical Sociology* (Collier–Macmillan, London).

COLLINS, N.R. and L.E.PRESTON, 1961, "The Size Structure of the Largest Industrial Farms, 1900–1958", *American Economic Review*, *51*, 986–1011.

COOTNER, P.H., 1964, *The Random Character of Stock Market Prices* (Massachusetts Institute of Technology Press, Cambridge).

CORNISH, E.A., 1954, "The Multivariate *t*-Distribution Associated with a Set of Normal Sample Deviates", *Australian Journal of Physics*, *7*, 531–542.

COX, D.R. and H.D.MILLER, 1965, *The Theory of Stochastic Processes* (Methuen, London).

CRAGG, J.G., 1964, "Small-Sample Properties of Various Simultaneous Equation Estimators: The Results of Some Monte Carlo Experiments", Econometric Research Program Research Memorandum No. 68, Princeton University.

CRAMER, H., 1946, Mathematical Methods of Statistics (Princeton University Press, Princeton) pp. 247–248.

DANTZIG, G.B. and A.ORDEN, 1953, "Duality Theorems", *RAND Report R.M.–1265*, Santa Monica, California, The RAND Corporation.

DEAN, G.W., S.S.JOHNSON and H.O.CARTER, 1963, "Supply Functions for Cotton in Imperical Valley, California", *Agricultural Economics Research*, *15*, 1–14.

DEUTSCH, R., 1965, *Estimation Theory* (Prentice-Hall, New York).

DORN, W.S., 1960, "Duality in Quadratic Programming", *Quarterly of Applied Mathematics*, *18*, 155–162.

DRYDEN, M.M., 1967, "Share Price Movements: A Markovian Approach", unpublished paper, Department of Economics, Edinburgh.

DUNNETT, C.S. and M.SOBEL, 1954, "A Bivariate Generalization of Student's *t*-Distribution, with Tables for Certain Special Cases", *Biometrika*, *41*, 153–169.

DURBIN, J., 1953, "A Note on Regression When There is Extraneous Information About One of the Coefficients", *Journal of the American Statistical Association*, *48*, 700–708.

EDWARDS, W., H.LINDMAN and L.J.SAVAGE, 1963, "Bayesian Statistical Inference for Psychological Research", *Psychological Review*, *70*, 193–242.

FARRIS, P.L. and S.I.PADBERG, 1964, "Measures of Market Structure Change in the Florida Fresh Citrus Packing Industry", *Agricultural Economics Research*, *16*, 93–102.

FERGUSON, T.S., 1967, *Mathematical Statistics: A Decision Theoretic Approach* (Academic Press, New York).

FISHER, W.D., 1961, "A Note on Curve Fitting with Minimum Deviations by Linear Programming", *Journal of the American Statistical Association, 56*, 359–363.

FRANK, M. and P.WOLFE, 1956, "An Algorithm for Quadratic Programming", *Naval Research Logistic Quarterly, 3*, 97–98.

GALE, D., 1960, *The Theory of Linear Economic Models*(McGraw–Hill, New York) pp.260–290.

GOLDBERGER, A.S., 1964, *Econometric Theory* (John Wiley and Sons, New York) pp.246–248.

GOODMAN, L.A., 1953, "A Further Note on Miller's 'Finite Markov Processes in Psychology'", *Psychometrika, 18*, 245–248.

GOODMAN, L.A., 1965, "On Statistical Analysis of Mobility Tables", *The American Journal of Sociology, 70*, 564–585.

HADLEY, G., 1964, *Nonlinear and Dynamic Programming* (Addison–Wesley Publishing Company, Boston) pp.212–250.

HART, P.E. and S.J.PRAIS, 1956, "The Analysis of Business Concentration, A Statistical Approach", *Journal of the Royal Statistical Society* (A), *119*, 150–175.

HARTLEY, H.O., 1963, "The Application of Mathematical Programming to Statistical Least Squares Estimation with Constraints", unpublished mimeographed report, University of Texas.

HAWKINS, D. and H.A.SIMON, 1949, "Note: Some Conditions on Macro Economic Stability", *Econometrica, 17*, 245–248.

HEADY, E.O. and W.CANDLER, 1963, *Linear Programming Methods* (The Iowa State University Press, Ames) pp.98–99; 131; 136.

HOCKING, R.R., 1963, "The Distribution of a Projected Least Squares Estimator", *Annals of the Institute of Statistical Mathematics, 17*, 357–362.

HOEL, P.E., 1954, "A Test for Markov Chains", *Biometrika, 41*, 430–433.

HOGG, R. and A.CRAGG, 1965, *Introduction to Mathematical Statistics*, 2nd Edition (Macmillan, New York).

HOWARD, R.A., 1960, *Dynamic Programming and Markov Processes* (John Wiley and Sons, New York).

JEFFREYS, H., 1961, *Theory of Probability*, 3rd Edition (Clarendon Press, Oxford).

JOHNSTON, J., 1963, *Econometric Methods* (McGraw–Hill, New York).

JUDGE, G.G. and E.R.SWANSON, 1962, "Markov Chains: Basic Concepts and Suggested Uses in Agricultural Economics", *Australian Journal of Agricultural Economics, 6*, 49–61.

JUDGE, G.G. and T.TAKAYAMA, 1966, "Inequality Restrictions in Regression Analysis", *Journal of the American Statistical Association, 61*, 166–181.

KAO, R.C.W., 1953, "Note on Miller's 'Finite Markov Processes in Psychology'", *Psychometrika, 18*, 241–243.

KARLIN, S., 1959, *Mathematical Methods and Theory in Games, Programming and Economics* (Addison–Wesley, London) pp.241–243.

KARST, O.J., 1958, "Linear Curve Fitting Using Least Deviations", *Journal of the American Statistical Association*, *53*, 118–132.

KEMENY, J.G., H. MIRKIL, J.L. SNELL and G.L. THOMPSON, 1959, *Finite Mathematical Structures* (Prentice-Hall, New York) pp. 348–438.

KEMENY, J.G. and J.L. SNELL, 1960, *Finite Markov Chains* (De van Nostrand Co., Inc., Princeton) pp. 200–206.

KENDALL, M.G. and A. STUART, 1958, *The Advanced Theory of Statistics*, Vol. 1 (Charles Griffin and Co., Ltd., London).

KENDALL, M.G. and A. STUART, 1961, *The Advanced Theory of Statistics*, Vol. 2 (Charles Griffin and Co., Ltd., London).

KISLEV, Y. and A. AMAID, 1968, "Linear and Dynamic Programming in Markov Chains", *American Journal of Agricultural Economics*, *50*, 111–129.

KOTTKE, M.W., 1964, *Patterns of Dairy Farm Exit and Growth*, Connecticut Agricultural Experiment Station Bulletin No. 382.

KRENZ, R.D., 1964, "Projection of Farm Numbers for North Dakota with Markov Chains", *Agricultural Economics Research*, *16*, 77–83.

KUHN, H. and A. TUCKER, 1951, "Non Linear Programming" in: J. Neyman, ed., *Proceedings of the Second Berkeley Symposium* (The University of California Press, Berkeley) pp. 481–492.

LEE, T.C., G.G. JUDGE and R.L. CAIN, 1969, "A Sampling Study of the Properties of Estimators of Transition Probabilities", *Management Science* (A), *15*, 374 to 398.

LEE, T.C., G.G. JUDGE and T. TAKAYAMA, 1965, "On Estimating the Transition Probabilities of a Markov Process", *Journal of Farm Economics*, *47*, 742–762.

LEE, T.C., G.G. JUDGE and A. ZELLNER, 1968, "Maximum Likelihood and Bayesian Estimation of Transition Probabilities", *Journal of the American Statistical Association*, *63*, 1162–1179.

LINDGREN, B.W., 1962, *Statistical Theory* (MacMillan Company, New York) pp. 300–304.

LINDLEY, D.V., 1965, *Introduction to Probability and Statistics*, Parts I and II (Cambridge University Press, Cambridge).

MADANSKY, A., 1959, "Least Squares Estimation in Finite Markov Processes", *Psychometrika*, *24*, 137–144.

MARSCHAK, J., 1950, "Statistical Inference in Economics", in: T.C. Koopmans, ed., *Studies in Econometric Method* (John Wiley and Sons, New York).

MARSCHAK, J., 1953, "Economic Measurements for Policy and Prediction", in: W.C. Hood and T.C. Koopmans, eds., *Studies in Econometric Method* (John Wiley and Sons, New York) pp. 10–26.

MARTIN, J.J., 1967, *Bayesian Decision Problems and Markov Chains* (John Wiley and Sons, New York).

MATRAS, J., 1960, "Differential Fertility, Intergenerational Occupational Mobility and Change in Occupational Distribution", *Population Studies*, *15*, 187–197.

MAULDON, J.G., 1961, "A Generalization of the Beta-Distribution", *Annals of Mathematical Statistics*, *32*, 509–520.

MEYER, J. and R. G. GLAUBER, 1964, *Investment Decision, Economic Forecasting and Public Policy* (Harvard University, Boston).

MILLER, G. A., 1952, "Finite Markov Processes in Psychology", *Psychometrika, 17*, 149–167.

MOSES, L. E. and R. V. OAKFORD, 1963, *Tables of Random Permutations* (Stanford University Press, Stanford) pp. 67–77; 190–229.

NEUDECKER, H., 1966, "A Critique of the Charnes–Cooper Procedure of Local Aggregation in Input–Output Analysis", Discussion paper 74, University of Birmingham.

NEYMAN, J., 1949, "Contribution to the Theory of the χ^2 Test", *The First Proceedings of Berkeley Symposium in Math. Stat. and Prob.*, 239 (University of California Press, Berkeley).

NICHOLLS, W. H., 1951, *Price Policies in the Cigarette Industry* (The Vanderbilt University Press, Nashville).

ODELL, P. L. and T. O. LEWIS, 1966, "A Generalization of the Gauss–Markov Theorem", *Journal of the American Statistical Association, 61*, 1063–1066.

PADBERG, D. I., 1962, "The Use of Markov Processes in Measuring Changes in Market Structure", *Journal of Farm Economics, 44*, 189–199.

PANNE, VAN DE and A. WHINSTON, 1963, "The Simplex and the Dual Method for Quadratic Programming", Research Report, International Center for Management Science, Rotterdam.

PENROSE, R., 1955, "A Generalized Inverse for Matrices", *Proceedings of the Cambridge Philosophical Society, 51*, 406–413.

PRAIS, S. J., 1955, "Measuring Social Mobility", *Journal of the Royal Statistical Society* (A), *118*, 56–66.

PRESTON, L. E. and E. J. BELL, 1961, "The Statistical Analysis of Industry Structure: An Application to the Food Industry", *Journal of the American Statistical Association, 56*, 925–932.

PRICE, C. M., 1964, "The Matrix Pseudo Inverse and Minimal Variance Estimates", *SIAM Review, 6*, 115–120.

RADO, R., 1956, "Note on Generalized Inverse of Matrices", *Proceedings of the Cambridge Philosophical Society, 52*, 600–601.

RAO, C. R., 1962, "A Note on the Generalized Inverse of a Matrix with Applications to Problems in Mathematical Statistics", *Journal of the Royal Statistical Society* (B), *24*, 152–158.

REITZ, H. L., et al., 1934, *Handbook of Mathematical Statistics* (Houghtton Mifflin Co., Boston).

REY, G., 1964, "A Markov Chain Prediction of Value Added and Expenditure Shares, Italy, 1861–1956", Research Report, International Center for Management Science, Rotterdam.

ROBERTS, H. V., 1963, "Bayesian Inference", *American Statistical Association, Proceedings of the Social Statistical Section*, pp. 76–80.

SAATY, T. L. and J. BRAM, 1964, *Nonlinear Mathematics* (McGraw-Hill Book Company) pp. 113–133.

SAMUELSON, P.A., 1947, *Foundations of Economic Analysis* (Harvard University Press, Cambridge) pp. 348–349.

SAVAGE, L.J., 1962, "Bayesian Statistics", in *Recent Development in Decision and Information Processes* (Macmillan and Co., New York) pp. 161–194.

SHUPP, F., 1968, "Stabilization Policies in Non-linear Stochastic Macro Economic Systems", unpublished paper, University of Illinois.

SIEGEL, S., 1956, *Nonparametric Statistics for the Behavior Sciences* (McGraw-Hill Book Company, New York) pp. 68–83; 229–238.

SMITH, P.E., 1961, "Markov Chains, Exchange Matrices and Regional Development", *Journal of Regional Science*, *3*, 27–36.

SOLOW, R., 1951, "Some Long-Run Aspects of the Distribution of Wage Incomes" *Econometrica*, *19*, 333–334.

SPARKS, W.R., 1960, "On Markov Chains in Demand Analysis", unpublished paper, Michigan State University.

STEINDL, J., 1965, *Random Processes and the Growth of Firms* (Hafner Publishing Company, New York).

SUMMERS, R., 1965, "A Capital Intensive Approach to the Small Sample Properties of Various Simultaneous Equation Estimators", *Econometrica*, *35*, 1–41.

TAIWAN PROVINCIAL GOVERNMENT, Department of Agriculture and Forestry, *Taiwan Agricultural Yearbook*, 1941–1967.

TAKAYAMA, T., G.G. JUDGE and T.C. LEE, 1969, "An Additional Note on Miller's 'Finite Markov Processes in Psychology'", Workshop Paper, University of Illinois.

TELSER, L.G., 1962a, "The Demand for Branded Goods as Estimated from Consumer Panel Data", *Review of Economic Statistics*, *44*, 300–324.

TELSER, L.G., 1962b, "Advertising and Cigarettes", *Journal of Political Economy*, *70*, 471–499.

TELSER, L.G., 1963, "Least Squares Estimates of Transition Probabilities", in: *Measurement of Economics* (Stanford University Press, Stanford).

THEIL, H., 1961, *Economic Forecasts and Policy*, 2nd Edition (North-Holland Publishing Company, Amsterdam) pp. 331–333.

THEIL, H., 1963, "On the Use of Incomplete Prior Information in Regression Analysis", *Journal of the American Statistical Association*, *58*, 401–414.

THEIL, H. and A.S. GOLDBERGER, 1961, "On Pure and Mixed Statistical Estimation in Economics", *International Economic Review*, *2*, 65–78.

THEIL, H. and GUIDO REY, 1966, "A Quadratic Programming Approach to the Estimation of Transition Probabilities", *Management Science*, *12*, 714–721.

THORNBER, H., 1967, "Finite Sample Monte Carlo Studies; An Auto-Regressive Illustration", *Journal of the American Statistical Association*, *62*, 801–818.

TIAO, G.C. and A. ZELLNER, 1964, "On the Bayesian Estimation of Multivariate Regression", *The Journal of the Royal Statistical Society* (B), *26*, 277–285.

TIAO, G.C. and A. ZELLNER, 1965, "Bayes' Theorem and the Use of Prior Knowledge in Regression Analysis", *Biometrika*, *51*, 219–230.

WAGNER, H.H., 1959, "Linear Programming for Regression Analysis", *Journal of the American Statistical Association, 54*, 206–212.

WOLFE, P., 1959, "The Simplex Method for Quadratic Programming", *Econometrica, 27*, 382–398.

ZELLNER, A., 1961, *Linear Regression With Inequality Constraints on the Coefficients*, Report 6109, Econometric Institute, Rotterdam.

ZELLNER, A., 1962, "An Efficient Method of Estimating Seemingly Unrelated Regressions and Tests for Aggregation Bias", *Journal of the American Statistical Association, 57*, 348–368.

ZELLNER, A., 1969, *Introduction to Bayesian Inference in Econometrics* (University of Chicago, Chicago) Manuscript.

ZELLNER, A. and T.H.LEE, 1965, "Joint Estimation of Relationships Involving Discrete Random Variables", *Econometrica, 33*, 382–394.

ZELLNER, A. and G.C.TIAO, 1964, "Bayesian Analysis of the Regression Model with Autocorrelated Errors", *Journal of the American Statistical Association, 59*, 763–778.

AUTHOR INDEX

Adelman, I.G., 18
Aitken, A.C., 63, 64, 79
Anderson, T.W., 18, 20, 23, 24, 25
Arrow, K.J., 18
Ashar, V.G., 135

Bailey, N.T.J., 20
Bartlett, M.S., 25
Bell, E.J., 18
Blackwell, D., 30
Blumen, I., 18
Boot, J.C.G., 165, 169
Bush, R.R., 18

Champernowne, D.G., 18
Charnes, A., 80
Chipman, J.C., 63
Coleman, J.S., 18
Cooper, W.W., 80
Cootner, P.H., 18
Cox, D.R., 17, 20
Cramer, H., 42

Dantzig, G.B., 40, 116
Deutsch, R., 166
Dorn, W.S., 40, 101

Ferguson, T.S., 26
Fisher, W.D., 98, 131

Gale, D., 18
Girschick, M.A., 30
Goldberger, A.S., 34, 113
Goodman, L.A., 18, 20, 23, 24, 25, 36, 39, 40

Hart, P.E., 18
Hartley, H.O., 42
Hawkins, D., 37
Hocking, R.R., 42
Hoel, P.E., 25
Howard, R.A., 18

Jeffreys, H., 112
Johnston, J., 57, 174
Judge, G.G., 34, 40

Kemeny, J.G., 20, 75
Kendall, M.G., 25, 50, 51, 53, 78, 83, 94, 98, 102, 141
Kuhn, H., 40, 80

Lee, T.C., 34, 40
Lee, T.H., 39, 78, 183, 190
Lindgren, B.W., 57
Lindley, D.V., 112

Madansky, A., 32, 38, 39, 63, 67
Marschak, J., 17
Martin, J.J., 28, 29, 30, 107, 108
Matras, J., 18
Mauldon, J.G., 28, 107
Miller, H.D., 17, 20, 32, 33, 40
Mosteller, F., 18

Neyman, J., 88
Nicholls, W.H., 60

Orden, A., 40, 116

Penrose, R., 165

251

Prais, S.J., 18
Preston, L.E., 18
Price, C.M., 165

Rao, C.R., 63, 165
Rado, R., 105
Reitz, H.L., 74, 93
Rey, G., 40, 67

Shupp, F., 18
Siegel, S., 50, 150, 154
Simon, H.A., 37
Smith, P.E., 18
Snell, J.L., 20
Solow, R., 18
Sparks, W.R., 18
Steindl, J., 18

Stuart, A., 51, 53, 78, 93, 94, 98, 102, 141

Takayama, T., 34, 40
Telser, L.G., 18, 36, 39, 40, 59, 99, 177, 191
Theil, H., 34, 40, 41, 59, 67, 113, 167
Thornber, H., 43
Tiao, G.C., 98, 113
Tucker, A., 40, 80

Wagner, H.H., 131
Wallace, T.D., 135
Wolfe, P., 41

Zellner, A., 34, 39, 41, 42, 78, 98, 112, 113, 183, 190

SUBJECT INDEX

Aggregate data, 20
Aggregate mean square error, 148
Aitken estimator, 79
Aperiodic, 19
Asymptotic theory, 25
Asymptotically normally distributed, 25

Bayes' theorem, 26, 27
Bayesian inference, 27
Best asymptotically normal estimator, 88
Beta distribution
  univariate, 28, 111
  multivariate, 28, 107, 123
Binomial test, 50

Characteristic roots, 165, 167, 169
Characteristic vector, 165, 167, 169
Chi-square
  chi-square test, 51, 52, 139
  minimum chi-square estimator, 85
  modified minimum chi-square estimator, 87
Coefficient of concordance, 150, 152, 154
Conditional probability, 31
Conditional probability density function, 19
Consistent estimator, 38
Contingency tables, 25
Control variables, 17

Diagonal matrix, 33
Disturbance covariance matrix, 63, 73, 74

Dual problem, 40
Dynamic, 17

Economic variables, 17
Equilibrium vector, 19
Ergodic, 19, 23

Gamma function, 28, 107
General linear probability model, 183
Generalized inverse, 163–173
Generalized least squares, 73
Goodness-of-fit test, 139

Hawkins–Simons condition, 37
Heteroscedasticity, 39, 63

Information matrix, 98
Initial state, 19
Inverse probability (see Bayes' theorem)
Irreducible, 19

Kolmogorov–Smirnov D Statistic, 57, 58
Kronecker expansion, 76, 77

Lagrangean, 24
Least squares estimator, 32
Leptokurtic prior, 121
Lexis scheme, 74, 93
Likelihood function, 27, 29, 95, 109
Likelihood ratio test, 25
Linear dependence, 33
Linear programming, 40, 131
Loss function, 30, 43, 118

Macro Bayesian estimator, 115

253

Macro data, 20, 31
Macro maximum likelihood estimator, 93, 99
Market shares, 59
Markov process, 17
Matched pairs test, 50
Matrix beta, 28
Mean absolute error, 50
Mean square error, 50
Micro Bayesian estimator, 25
Micro maximum likelihood estimator, 23
Micro units, 20, 23
Minimum absolute deviations estimator, 131
Minimum chi-square estimator, 86
Mode of posterior distribution, 110
Modified minimum chi-square estimator, 87
Multinomial distribution, 24, 73, 74, 85, 93
Multinomial process, 109
Multivariate beta PDF, 28, 101, 108, 123
Multivariate linear statistical model, 33

Non-negative condition, 34
Non-parametric tests, 50
Non-spherical disturbances, 75
Normality test, 57, 58, 70

Optimal point estimate, 117, 118
Orthogonal matrix, 165

Pairwise comparisons, 50, 150
Parameter space, 26
Platykurtic prior, 125
Poisson scheme, 185
Positive definite matrix, 33
Posterior distribution, 109, 110
Predicted proportions, 139
Primal problem, 40
Primal–dual problem 41
Principle of stable estimation 127
Prior probability density function, 27

Prior distribution, 107
Probability limit, 38, 49
Probability vector, 44
Programming tableau, 41
Proportions, 31

Quadratic programming, 41, 80, 89, 101, 116

Random variables, 17, 18
Reduced model, 75, 76, 174, 176
Reduced weight matrix, 174
Restricted least squares estimates, 58
Restricted least squares estimator, 39, 40, 41
Risk function, 43
Root mean square error, 49, 145
Row sum condition, 34

Sample proportions, 47
Sampling experiment, 43
Sampling properties, 41
Signed rank rest, 150
Simulated population, 46
Simulated probability model, 44
Simulation, 46
Simultaneous, 17
Singular covariance matrix, 33
Slack variables, 41, 81, 101, 116
Stationary or equilibrium vector, 19, 20
Stationary Markov process, 23
Stochastic, 17
Stochastic process, 17, 18

Tests of fit, 53
Transition probabilities, 19

Unbiased estimator, 49
Univariate beta PDF, 38, 111
Unrestricted least squares estimates, 56
Unrestricted least squares estimator, 32

Variable transition probabilities, 191

Weighted least squares , 64